Endorsed by
University of Cambridge International Examinations

Complete Biology for Cambridge IGCSE

Teacher's Resource Kit

Ron Pickering

OXFORD UNIVERSITY PRESS

Great Clarendon Street, Oxford OX2 6DP

Oxford University Press is a department of the University of Oxford. It furthers the University's objective of excellence in research, scholarship, and education by publishing worldwide in

Oxford New York

Auckland Cape Town Dar es Salaam Hong Kong Karachi Kuala Lumpur Madrid Melbourne Mexico City Nairobi New Delhi Shanghai Taipei Toronto

With offices in

Argentina Austria Brazil Chile Czech Republic France Greece Guatemala Hungary Italy Japan Poland Portugal Singapore South Korea Switzerland Thailand Turkey Ukraine Vietnam

First published 2011

British Library Cataloguing in Publication Data

Data available

ISBN: 978-0-19-913879-1

10 9 8 7 6 5 4

Printed in Great Britain by WM Print

Paper used in the production of this book is a natural, recyclable product made
from wood grown in sustainable forests. The manufacturing process conforms to the environmental regulations to the country of origin.

Acknowledgments
The publishers would like to thank the University of Cambridge Local Examinations Syndicate for kind permission to reproduce past paper questions.

The University of Cambridge Local Examinations Syndicate bears no responsibility for the example answers to questions taken from its past question papers which are contained in this publication.

Cover photo by Gary Yim/Shutterstock

Contents

Practical activities

Investigations

Answers

Introduction to Teacher's Resource Kit for Cambridge IGCSE Biology

This book is written for teachers preparing students for the 2012 IGCSE Biology (syllabus code 0610) qualification from the University of Cambridge International Examinations, and complements the material presented in the student's book, *Complete Biology for Cambridge IGCSE.*

This book contains a number of resources which will enable the teacher to deliver the course more easily and effectively:

A series of practical exercises This series of exercises provides guidance for practical work which might be used to support the content of the student's book. Each exercise includes a list of materials and apparatus to be used, and step-by-step instructions on the collection of valid data. Materials and apparatus are chosen to be simple and readily available in most centres delivering this subject. Exercises are quantitative wherever possible, and each of them includes appropriate assessment opportunities.

A series of suggested investigations These are longer exercises, designed to guide the student through the principles of scientific investigation. The basic content will be familiar from work covered in the student's book, but the context may be made more challenging. Students will have the opportunity to collect data, and present it in appropriate tabular and graphical format. Throughout, the need to communicate information simply and effectively is stressed.

Answers to questions embedded in the student's book These include suggested answers to the questions associated with the topic 'spreads' in the student's book, and should prove valuable in assessing the accuracy of students' answers to work set for preparation or homework purposes. The answers also include those for the past paper questions used for summative assessment, ensuring that students are working to the standards required by the end-of-course examinations.

The CD associated with the book includes a series of **additional activities**. These can be set as homework, or as extensions for the more able students. These activities put content from the student's book into a more applied setting, and will be valuable in emphasising 'How Science Works'.

Ron Pickering
June, 2010

 Name:

Microscopes and how to use them

You need:

- a microscope
- prepared slides (e.g. insect head, wings, legs etc.)

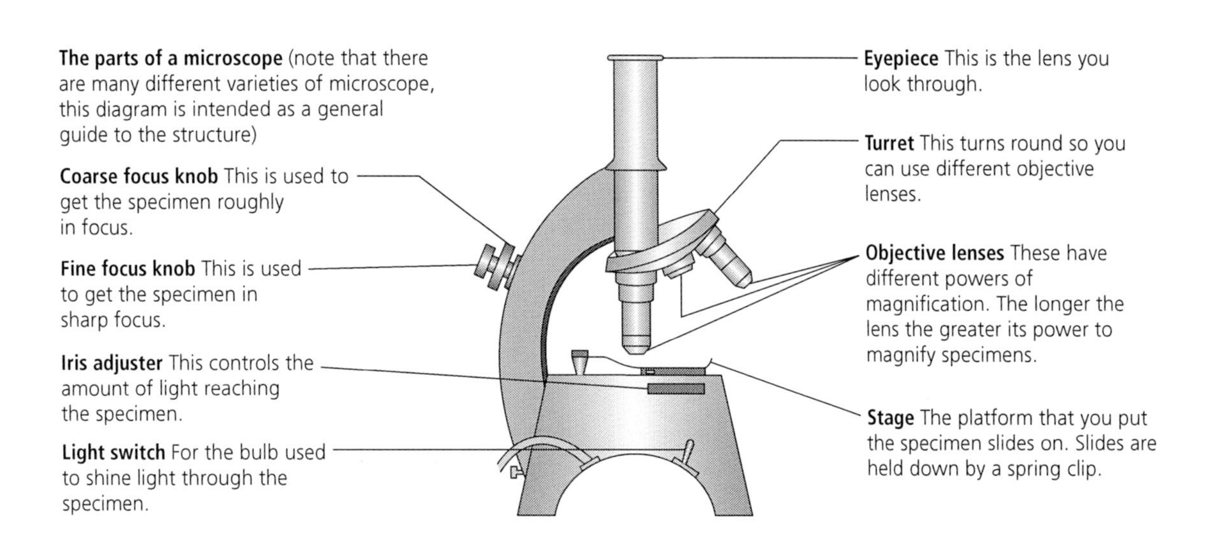

Using a microscope

1. Turn the turret until you have the low power objective lens (the short lens) in line with the eyepiece.
2. Clip a slide on the stage so that it is in the centre under the objective lens and look through the eyepiece.
3. Adjust the coarse focus until the specimen becomes clear. If necessary adjust the fine focus until the specimen is in sharp focus.
4. Move the iris adjuster until the specimen is clearly lit.
5. Calculate the magnification by multiplying the power of the eyepiece by the power of the objective lens (e.g. a $\times$ 5 eyepiece used with a $\times$ 15 objective magnifies 75 times).
6. Notice how, when you move the slide, the specimen seems to move in the opposite direction.
7. Change to the medium power objective. **Do not use the higher power objective yet.** Focus the microscope and notice that you now see much less of the specimen but at a higher magnification.
8. **How to use high power objectives.** If this is done carelessly the lens and a slide can be damaged. With some microscopes, equal damage can be done with smaller lenses too.
   a) With your eyes level with the stage, slowly lower the high power objective until it *almost* touches the slide.

 Name: ....

b) Look through the eyepiece and focus by moving the lens *away from the slide* (i.e. always focus upwards). This avoids smashing the lens through the slide.

Care of microscopes

When not in use, keep the microscope protected with a plastic cover (and in a box if possible).

Accumulated dust in a microscope can deteriorate image quality. Keep all openings covered with dust caps so that dust does not enter the microscope and settle on inaccessible lenses, mirrors, and prisms.

Use an air blower to blow dust off the stage, base, and body. If necessary wipe it down with a damp cloth, and clean off any smears with ethanol. Take care to clean and wash off the stage if any corrosive substance (even a salt solution) has been used.

Carefully clean the objectives (see detailed description below).

Remove the condenser top lens and clean it with lens paper and ethanol, if necessary.

Remove both eyepieces and clean their surfaces with an alcohol swab/lens tissue. Blow any dust or dirt out of the insides with an air blower, if one is available.

An occasional thorough cleaning of immersion objectives is necessary, but try to avoid doing this too often (more than once per month), as cleaning agents can remove an objective's anti-reflection coating over a period of time.

To clean a lens, remove it from the turret. Fold a piece of lens tissue into quarters, and add a few drops of straight ethanol. Gently wipe the lens in a circular motion (only letting the tissue, not your fingers, come into contact with the lens glass). Always immediately wipe off any excess ethanol with a dry piece of tissue – allowing ethanol to remain on the lens could also affect the anti-reflection coating (and may slowly loosen the cement which holds the lens in position).

Examine the lens carefully by removing the microscope's eyepiece, looking through it backwards, to see a magnified image of the lens. The lens surface should appear spotlessly clean. If not, repeat the above procedure. This is also a good way to examine a lens closely for scratches or other imperfections.

Your teacher will be looking for:

- *careful* use of the apparatus given
- good observation of the point where the image is in sharp focus

Name: ....

Making microscope slides

You need:

- microscopes
- mounted needles
- slides and coverslips
- scissors
- pond or aquarium water
- newspaper with words and pictures
- crystals (salt, sugar, potassium manganate(VII), copper sulfate)
- dropper pipettes

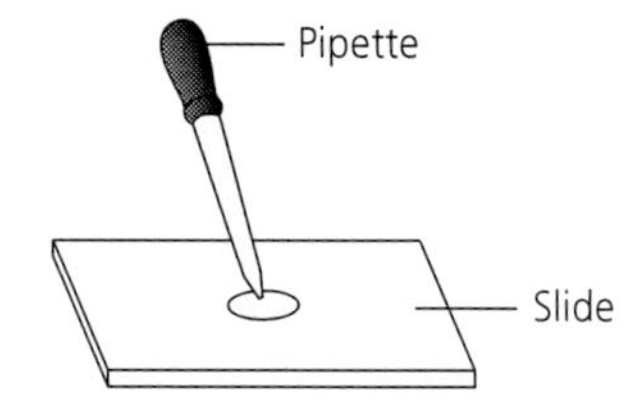

Pond and aquarium water

Water from the bottom of a pond or aquarium, especially if it contains rotting vegetatian, can contain many different protozoa and other microscopic creatures.

1 Use a bulb pipette to place one drop of a pond or aquarium water onto the centre of a glass slide.

2 Place a coverslip with one edge resting on the slide near the drop of water. Use a mounted needle to lower it slowly onto the water. If you do this quickly you will trap air bubbles. Use just enough water to spread to the edges of the coverslip and no further. Place the slide on the microscope stage.

3 Start with low power magnification and search the slide for interesting objects, then change to medium or high power magnification.

4 Make notes and drawings of what you find.

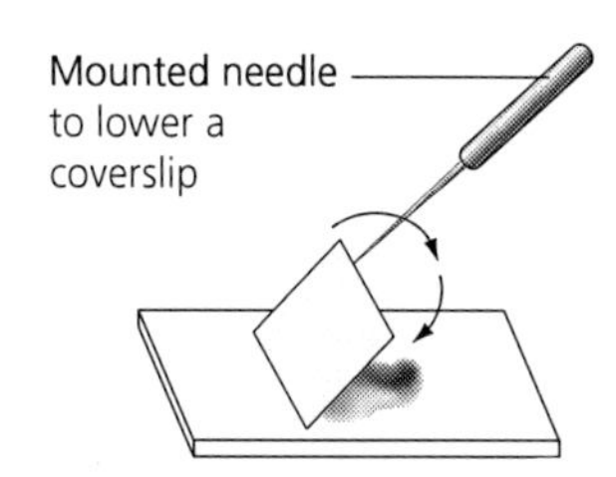

More things to do

5 Put a drop of tap water onto a slide. Remove a hair from your head, place it across the water and lower a coverslip over it. Study it under medium and high power magnification and make notes and drawings of the root end, the middle and the upper end of the hair.

6 Cut out pieces of newspaper small enough to fit under a coverslip. Mount them in water on slides. What is a newspaper photograph made up of?

7 Sprinkle some crystals on a *dry* slide. Study them without a coverslip. Prepare a table and use drawings and words to compare the shape and colour of four different types of crystal.

Correctly prepared slide

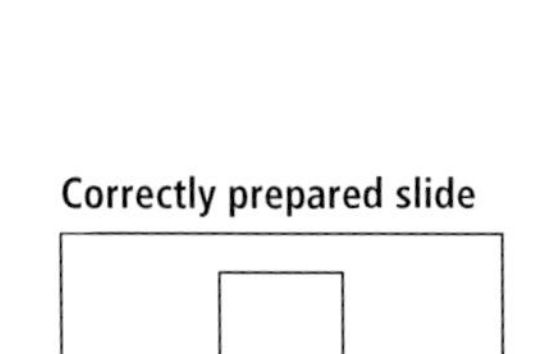

Badly prepared slide

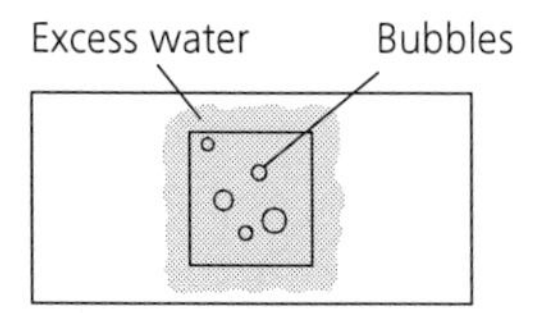

HAZARD WARNING

Copper sulfate is harmful when swallowed. It may also be irritating to eyes and skin. Potassium manganate VII is harmful if swallowed. AVOID SKIN CONTACT. WEAR EYE PROTECTION.

Your teacher will be looking for:

- careful use of the apparatus given
- good observation
- good presentation of results through diagrams

PRACTICAL **Name:**

Looking at cells

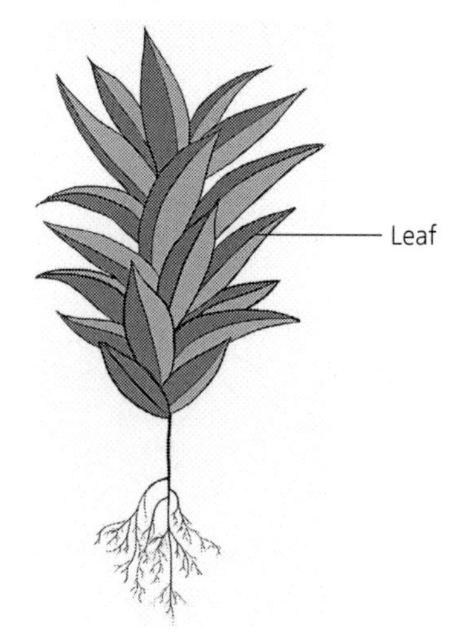

You need:

- microscopes
- razors or scalpels
- slides and coverslips
- Petri dishes
- forceps
- onions
- kidneys
- moss plants

Moss leaf cells

1 Use forceps to take one leaf off a moss plant. Put the leaf on a slide, add a drop of water and lower a coverslip onto it.

2 Observe it under low, medium and high power. Identify as many parts as you can.

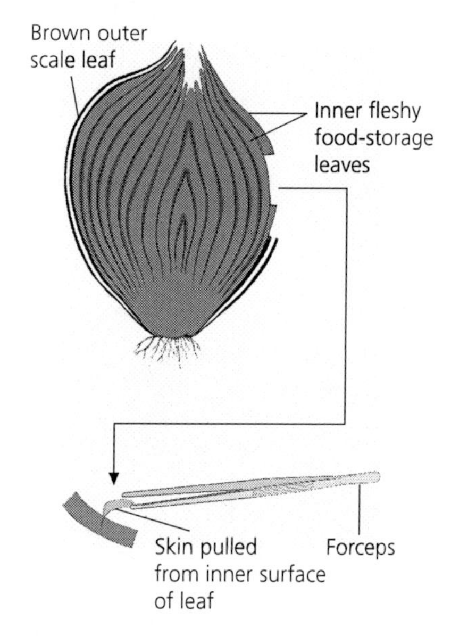

Onion cells

3 If you look at half an onion, you will see that it is made of fleshy leaves. Use a razor to cut a small piece out of one of the leaves. Use forceps to peel skin off the *inner* surface of the leaf. This skin is a thin layer of living cells. Put the skin into a Petri dish of water. It is important to cut a small piece of skin (less than 5 mm), as a larger piece will keep the curvature of the onion and will not stay flat on a slide.

4 Put a drop of iodine stain onto a slide. Put a piece of onion skin into the stain and smooth it out so there are no folds. Lower a coverslip over it, taking care not to trap any bubbles. The bubbles will look like perfectly circular car tyres. Prepare another slide in the same way but using water instead of iodine stain.

5 Study the stained onion cells under different magnifications, then look at unstained cells. What parts of the cells have become stained? How are onion cells *different from*, and *similar to*, moss leaf cells?

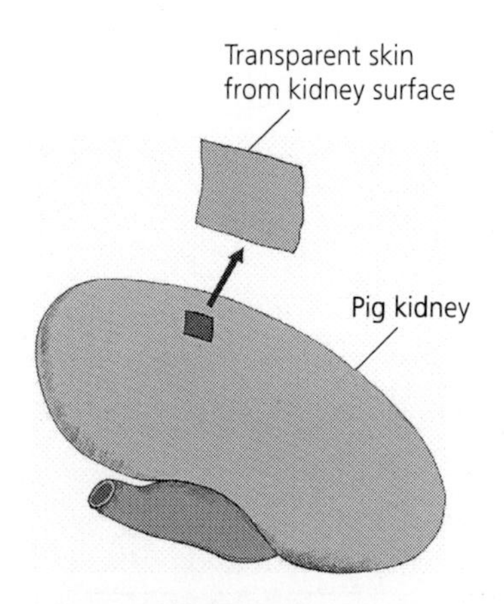

Animal cells

6 Use a razor and forceps to peel small pieces of transparent skin off the outside of a kidney. Make a slide of the skin in water, and another in iodine.

7 Study stained and unstained cells. How are they different? Draw moss leaf, onion and animal cells and list their *similarities* and *differences*.

Your teacher will be looking for:

- careful and skilful use of the apparatus given
- good observation of cell structure
- good presentation of results including drawings

HAZARD WARNING

Scalpels or razors are sharp, handle with care.

Name: ....

Measuring cells

You need:

- microscopes
- slides and coverslips
- razor blades
- Petri dishes
- onions and moss plants
- clear plastic rulers
- salt and other crystals
- insect slides (permanent)

Measuring a field of view

1 Place a clear plastic ruler under a microscope and focus on it with low power magnification. How many millimetres wide is the field of view?

2 **Problem:** Microscopic objects are measured in **micrometres** (one micrometre is written 1μm). 1mm = 1000μm. Convert your field of view to micrometres.

Ruler

Millimetre marks

Measuring onion cells

3 Prepare a slide of onion cells. Look at the slide under low power magnification. How many cells fit across the field of vision? In the drawing opposite, four and a half cells fill a field of view 2200μm wide. What is average length of each cell?

4 What is the average length, in micrometres, of onion cells in your slide? Turn the slide around and calculate the average width of the cells.

5 You now know the length in micrometres of one onion cell. Use this information, and your onion slide, to calculate the field of vision in micrometres under medium and high power magnification.

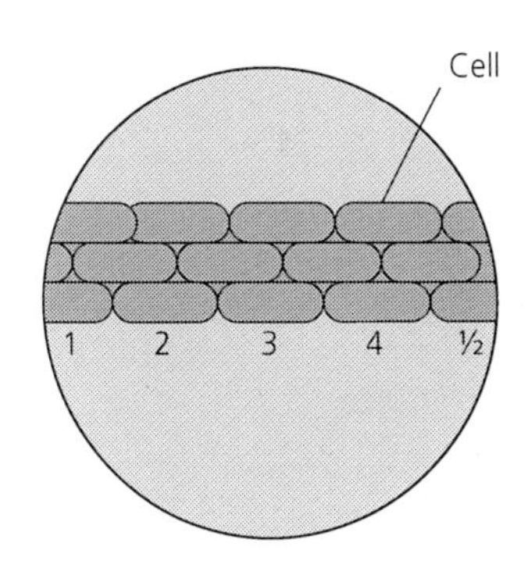

More things to do

6 Using the technique you have learned, measure:

a) the length and width of moss leaf cell

b) the width of a human hair

c) the average size of sugar, salt and other crystals

7 Look at permanent slides of insects and measure various parts, such as the width of scales on a butterfly's wing, the width of lenses in an insect's compound eye, the size of a fly's foot, etc.

Your teacher will be looking for:

- careful use of the apparatus given
- accurate measurements and calculations
- good presentation of results

HAZARD WARNING

Razor blades are sharp, handle with care.

PRACTICAL **Name:**

The response of small invertebrates to moisture

You need:

- shallow plastic or enamel dish
- cotton wool or wood shavings
- fine paintbrush
- stopclock or watch with 'second' hand
- marking pen
- small terrestrial invertebrates (e.g. woodlice or blowfly larvae)

Method

1 Use the marker pen to draw a line that divides the dish into two halves, and to mark one end of the dish 'dry' and the other end 'moist'.

2 Place a cotton wool ball at each end of the dish. Moisten the ball at the end of the dish marked 'moist' – it should be damp but there should be no excess water running from it.

3 Place 10 invertebrates in the dish, as close to the centre line as possible.

4 Cover the dish with a lightproof sheet e.g. cardboard.

5 After two minutes remove the cover. Count and record the number of animals at the 'dry' and 'moist' ends of the dish.

6 Use the paintbrush to gently move the animals back to the centre line, and repeat steps 3 to 5 until you have ten sets of counts.

Count	1	2	3	4	5	6	7	8	9	10	mean
Number on 'dry' side											
Number on 'moist' side											

Questions

a Why was the dish covered between readings?

b Was the 'dry' or 'moist' part of the dish favoured? How does this relate to the habitat in which these animals are normally found?

c The covering of woodlice or blowfly larvae is not waterproof. Explain exactly how these animals would lose water, and suggest what would happen to their cells if they did lose water.

d Why were ten animals used rather than just one, and why was the experiment repeated nine times?

PRACTICAL Name:

Making a key

You need:

- nails
- staples
- screws

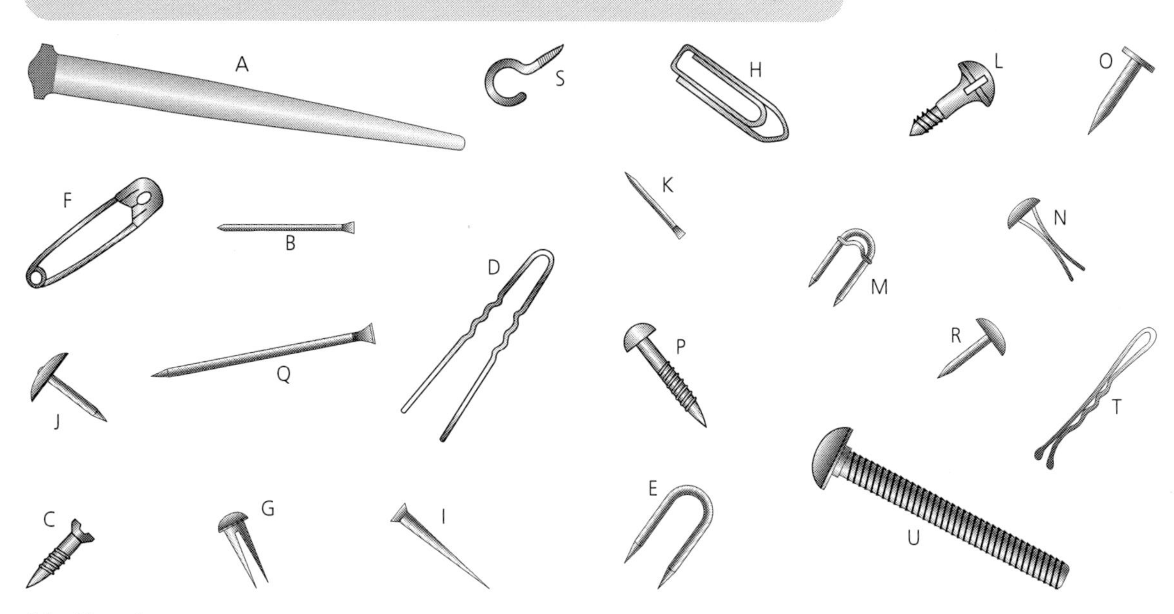

Method

1 Examine the collection of objects. Choose some feature that allows you to divide the group into two approximately equal-sized sub groups. Draw a 'branch/fork', and use the feature that you have chosen to place each object into one of the two groups. For example

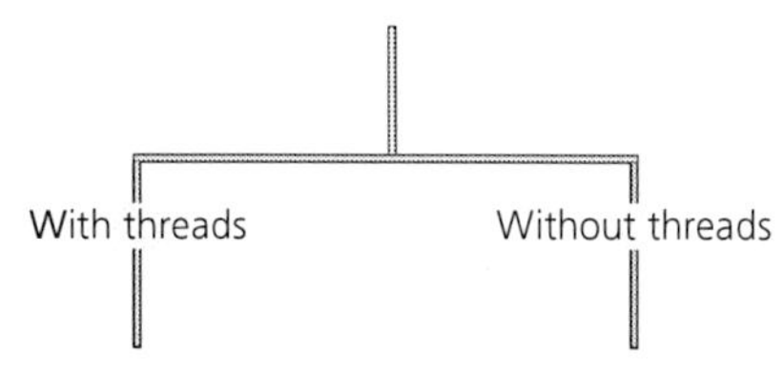

2 Repeat this branching process until you have separated all of the objects on the basis of some observable external characteristic.

3 Remove all of the objects, and ask one of your classmates to use the key to identify one of the objects (you choose which one they should try to identify).

Extension

Use the same objects to make an evolutionary tree. Place the simplest object first, and then make sub-groups that differ only in one characteristic from this simple 'ancestor'. You will end up with objects in the same group sharing many features, and they will probably be quite complex organisms.

What is the name of the classification group that contains only organisms that are so similar to one another that they can interbreed and produce fertile offspring?

PRACTICAL **Name:** ....

Osmosis

You need:

- potatoes
- Petri dishes
- dandelion/daisy stems
- test tubes and racks
- scalpels
- sugar
- razor blades

Osmosis in potato cells

1 Make three potato cups from raw potatoes cut in half: cut a depression in the top, peel skin off the sides, and give each a flat base (see diagram opposite).

2 Boil one cup. Place the cups in Petri dishes of water. Pour sugar into the boiled cup and into one of the raw potato cups. Leave one cup empty.

3 Observe and describe what happens to the sugar, and what happens in the empty cup.

Explain what happens in the three cups.

Why was one cup left empty?

What conclusions can you draw about osmosis in living and dead cells?

A Empty potato cup

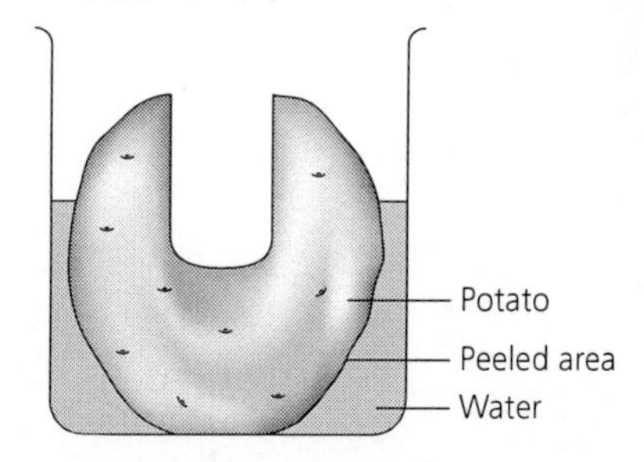

B Raw potato cup with sugar

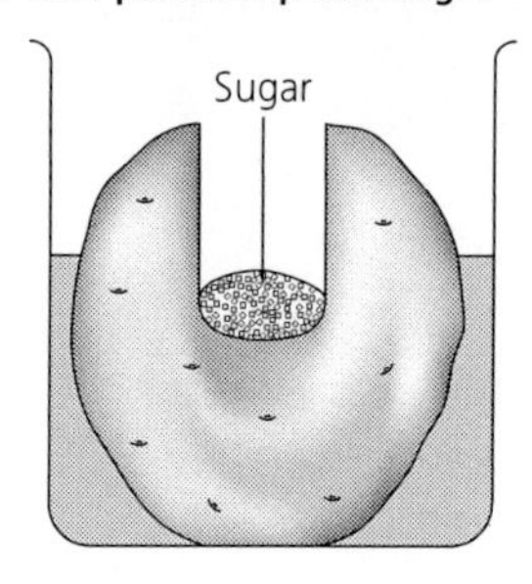

C Boiled potato cup with sugar

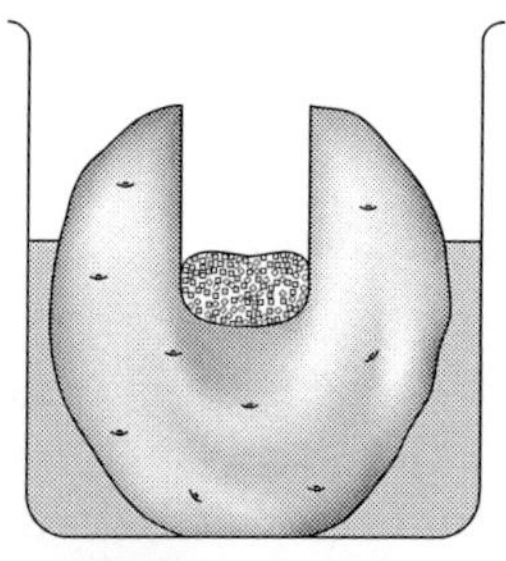

Osmosis in dandelion stalks

4 Take two test tubes. Half-fill one with water and the other with strong sugar solution.

5 Obtain two dandelion stalks. Slit them upwards for about 2.5 cm, then make a second upward slit at right angles to the first to divide the stalk base into four strips (see diagram).

6 Put one stalk in water and the other in sugar solution and leave them for 10 minutes.

7 Observe and describe what happens.

What happens if you move the stalk in water to the sugar solution and vice versa?

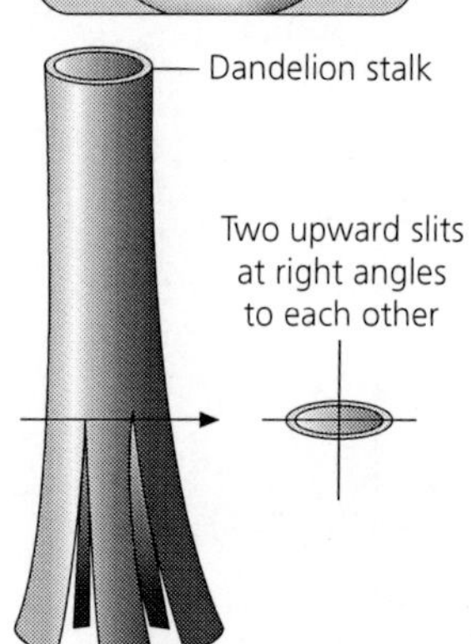

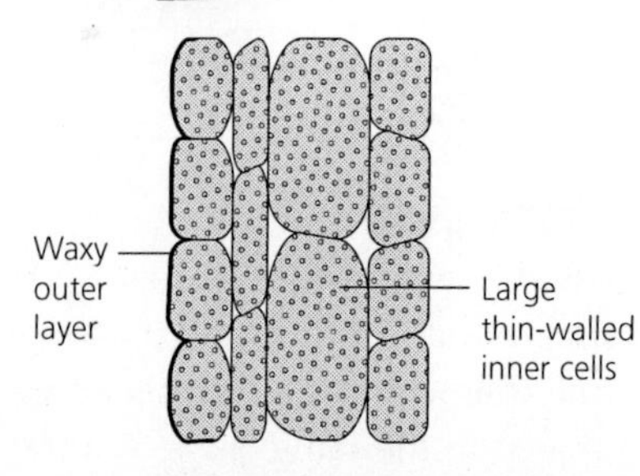

Your teacher will be looking for:

- skilful and safe use of the apparatus given
- good observation of results
- good presentation of results (including diagrams)
- sensible conclusions

Extension

The diagram opposite shows a section through a dandelion stalk. Study it and use it to form a hypothesis to explain what you observed in your experiment.

HAZARD WARNING

Razor blades are sharp, handle with care.

PRACTICAL **Name:** ....

Food tests

You need:

- iodine
- Benedict's reagent
- Biuret reagent
- ethanol
- test tubes and racks
- spotting tiles
- Bunsens, tripods and gauzes
- 250 cm^3 beakers
- goggles
- glass rods
- spatulas
- bulb pipettes
- sodium hydrogen carbonate
- liquid egg albumen
- bread
- potatoes
- cheese
- cooking oil
- glucose
- starch powder
- milk powder
- peanuts
- suet
- peas and beans
- carrots
- grapes

Begin by performing the following tests on known foods to observe a **positive result**. It is recommended that you then repeat each test with sodium bicarbonate to observe **negative result**. These observations will be helpful when you go on to test foods of unknown composition.

1 **Test for starch**
Place a little starch powder in a depression on a spotting tile.
Add a few drops of iodine.
Positive result: blue/black colour
Negative result: brown colour

2 **Test for glucose**
Place equal quantities of a strong glucose solution and Benedict's solution in a test tube (about 2 cm^3 of each).
Lower the test tube into a beaker of boiling water, wait until the test tube contents boil and leave it for two minutes.
Strong positive result: brick red precipitate
Medium positive result: yellow orange precipitate
Weak result: green colour
Negative result: blue colour
Before testing a solid food it must be crushed in warm water to extract any glucose which may be present.

3 **Test for proteins**
Dissolve a little milk powder in water in a test tube. Add a few drops of Biuret reagent.
Positive result: purple colour
Negative result: blue colour
Note that this is a test for soluble proteins. Before testing a solid food it must be crushed in warm water to dissolve any proteins which may be present.

HAZARD WARNING

Iodine and Benedict's are harmful to skin and eyes. AVOID SKIN CONTACT. WEAR EYE PROTECTION.

Food tests

4 **Test for oil and fat**
Place about 1 cm^3 of ethanol in a test tube. Add a few drops of oil and mix by shaking. Add an equal amount of water and shake again.
Positive result: a cloudy emulsion forms
Negative result: liquid remains clear
Food containing solid fats are tested by crushing them in ethanol to obtain an alcoholic solution. This is filtered and added to water.

More things to do

Use these tests to analyse the range of foods provided.
Divide each food into four samples and perform one test on each.
Remember to crush solid samples in warm water to extract glucose and protein, and alcohol to extract fats and oils.
Make sure you know the difference between positive and negative results.
Design a results table to show positive and negative results for each test.
List the types of food found in each sample.

From your tests list the types of food present in bread, milk, boiled potato and cheese.
Would eating these foods give you a balanced diet (i.e. do they contain sufficient carbohydrate, protein and fat for health)?
What would be the result of basing your diet on these foods alone?

HAZARD WARNING

Ethanol is highly flammable. KEEP AWAY from naked flame. Biuret is harmful to skin and eyes. AVOID SKIN CONTACT. WEAR EYE PROTECTION.

Your teacher will be looking for:

- careful and safe use of the apparatus given
- good observation
- good presentation of results

 Name: ....

Factors affecting the activity of catalase

Catalase is an enzyme found in many plant and animal tissues. The enzyme has the function of breaking down dangerous peroxide ions. As it does this, oxygen gas is released – because the reaction is rapid the oxygen bubbles form froth on the top of a reacting solution.

You need:

- 3 test tubes in a test tube rack
- hydrogen peroxide (20 volume) solution
- 10 cm^3 measuring cylinder
- forceps; dropper pipette
- boiling water bath
- watch glasses
- ruler (with mm scale)
- raw liver, cut into 5 g cubes
- pestle and mortar
- glass rod

Method

1 Label the test tubes A, B and C and the watch glasses B and C.

2 Measure out 5 cm^3 of hydrogen peroxide solution into each test tube.

3 Place one cube of raw liver into the boiling water bath, and leave for one minute.

4 Use the forceps to remove the cube from the water bath, and place it on watch glass B.

5 Grind one raw liver cube with the pestle and mortar, and transfer the paste to watch glass C.

6 The next step must be carried out quickly and very carefully. Add the remaining raw cube of liver to test tube A, the boiled cube to test tube B and the raw liver paste to tube C. You should try to add the liver to the test tubes as close to the same time as you possibly can.

7 After one minute, measure and record the height of froth in each of the tubes.

Tube	Treatment of liver	Height of froth after one minute (cm)
A	Raw cube	
B	Boiled cube	
C	Ground paste	

Try to explain your results. Try to use the terms *active site* and *denaturation* in your explanation.

Extension

Write out an equation for the reaction catalysed by catalase. Try to find out why peroxide ions are so dangerous to cells.

Try the same experiment, using potato cubes instead of liver. Explain any difference between the results for the animal and plant tissues.

PRACTICAL Name:

Measuring the energy values of foods

You need:

- Bunsen burner
- wood splints
- stands and clamps
- boiling tubes
- mounted needles
- thermometers
- safety goggles, screen
- measuring cylinder
- foods: pasta shape, sunflower seeds, bread

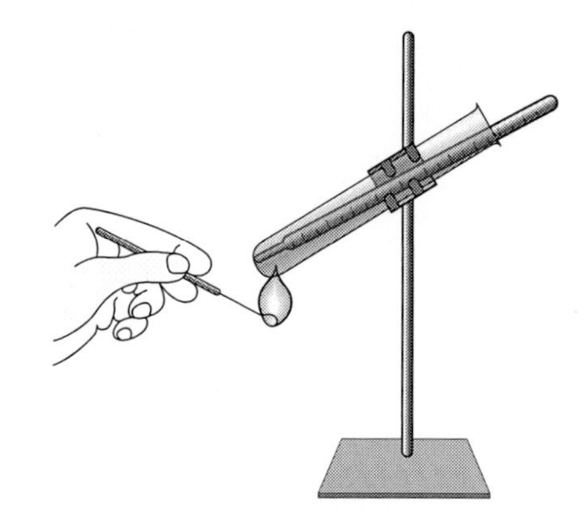

1. Put 20 cm^3 of water into a boiling tube. Fix the tube in a clamp so that it is held at an angle of 45° (see diagram).
2. Weigh a pasta shape very carefully, in grams (if possible to two decimal places), and note the result.
3. Fix the pasta onto a mounted needle, taking care that no bits drop off.
4. Measure the temperature of the water in the boiling tube and note the result.
5. Ignite the pasta in a Bunsen burner flame. *Quickly* place the burning pasta under the boiling tube. The idea is to use as much heat as possible from the burning pasta to heat the water in the tube.

   If the pasta goes out, relight it quickly and put it back under the tube.

   When the pasta has completely burnt, measure the temperature of water in the boiling tube again and note the result.
6. Before you can go any further you must know:
   - the mass of water in the boiling tube (1 cm^3 of water weighs 1 g)
   - the rise in temperature of water in the boiling tube
   - the mass of the pasta
7. It takes 4.2 joules of energy to raise the temperature of 1 g of water by 1°C, therefore you can calculate the energy given off by 1 g of pasta as follows:

$$\frac{\text{mass of water (in grams)} \times \text{rise in temperature} \times 4.2}{\text{mass of the pasta}}$$

8. Compare your result with what is stated on the food packaging. Your result will be much lower than the actual energy value of 1 g of pasta. Give as many reasons as you can why this is so. Despite this fact, if you use this method to find out the energy value of other foods, your results can still be compared. Why is this so?
9. Use this method to find out the energy values (in joule per gram) of the foods provided. Produce a results table and comment on your findings.
10. Design an improve method which will give a more accurate result. (Hint: is there any way of reducing heat loss to the air?)

Your teacher will be looking for:

- careful use of the apparatus given
- accurate measurements of volume of water, mass of food, temperature
- accurate recording of results and successful calculations of energies
- critical evaluation of the experiment and sensible suggestions for improving it

HAZARD WARNING

Wear EYE PROTECTION when burning foods. Use a safety screen.

 Name: ....

Photosynthesis and oxygen

You need:

- a one-litre beaker
- glass funnel
- test tube
- plasticine
- wood splint
- Bunsen burner
- bench lamp
- sodium hydrogen carbonate
- spatula
- pond weed *(Elodea)*

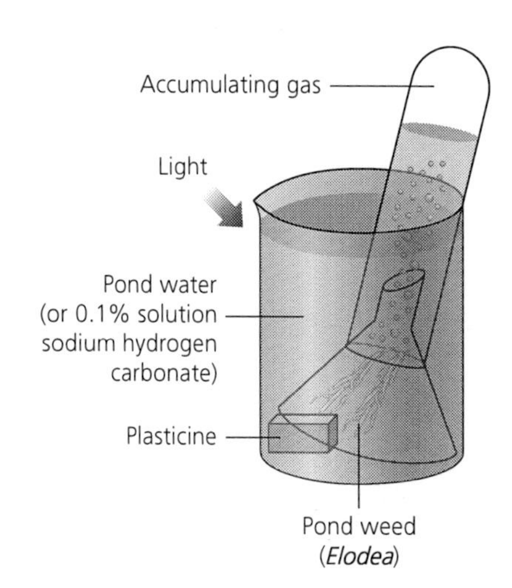

1. Three-quarters fill a one-litre beaker with water in which a small amount of sodium hydrogencarbonate has been dissolved. This will supply the plants with carbon dioxide.
2. Put a few springs of healthy pond weed such as *Elodea* in the bottom of the beaker. Place a glass funnel over the pond weed. Use one or more lumps of plasticine to raise the rim of the funnel off the base of the beaker, so the liquid can circulate freely. Make sure the liquid level is well above the end of the funnel.
3. Fill a test tube with weak sodium hydrogen carbonate solution. Put your thumb over the end, turn the tube upside down and lower it into the beaker without letting in any air. When the end of the test tube is under the liquid remove your thumb and lower the tube onto the funnel, as shown in the diagram.
4. Either put the apparatus on a well-lit window ledge or place it near a bench lamp.
5. After about a week sufficient gas should have collected in the tube to test.

   Lift the test tube off the funnel but do not let in any air. Put your thumb over the end of the tube, lift it out of the beaker and turn it right way up. Do not remove your thumb yet.

   Test the gas for the presence of oxygen: light a wood splint and when it is burning brightly blow it out so the end is glowing red hot. Lift your thumb off the test tube and *very quickly* lower the glowing wood splint into it. Observe closely what happens.

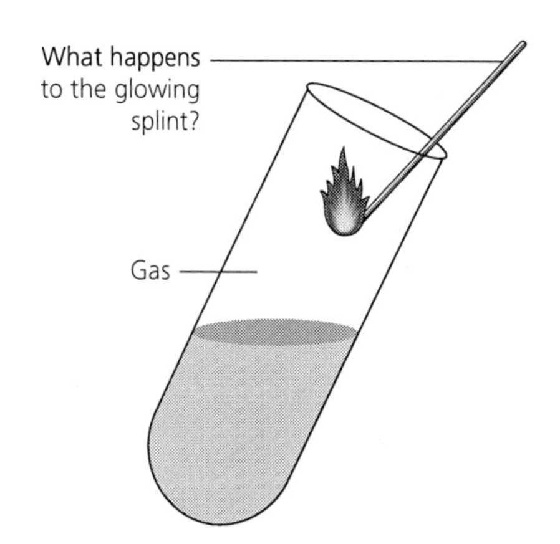

Questions

a Does the tube contain pure oxygen?

b Can you devise a control for this experiment?

Your teacher will be looking for:

- careful use of the apparatus given
- accurate observations
- good presentation of results
- sensible conclusions which fit your results

PRACTICAL **Name:**

Chlorophyll and photosynthesis

You need:

- plants with normal and variegated leaves (pelargonium, coleus or geranium)
- iodine solution
- boiling tubes
- tripods and gauzes
- forceps
- 500 cm^3 beakers
- white tiles
- Bunsen burner
- ethanol (alcohol)

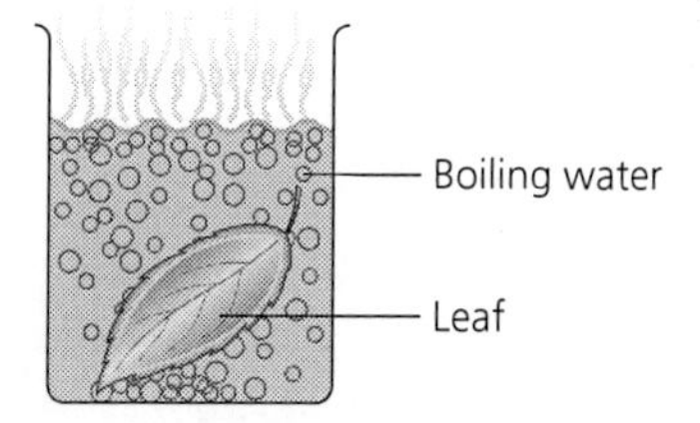

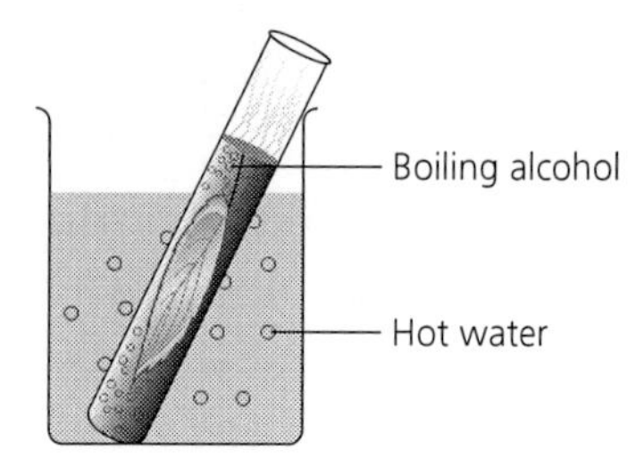

To show that plants need chlorophyll for photosynthesis you need a way of showing that photosynthesis has taken place. Plants change sugar produced by photosynthesis into starch and store it in their leaves. So a leaf with starch has been carrying out photosynthesis.

Testing a leaf for starch: method

1. Take a leaf from a non-variegated plant which has been in the light for a few hours. *Put goggles on.* Half fill a 500 cm^3 beaker with water and bring it to the boil. Put the leaf in the water for about 1 minute, then turn the Bunsen burner off.
2. Half fill a boiling tube with ethanol. *This is highly inflammable so do not put it near a naked flame.* Use forceps to take the boiled leaf out of the water and transfer it to the ethanol. Put the tube of ethanol into the beaker of very hot water. The ethanol will boil and remove chlorophyll from the leaf, making test results easier to see.
3. Lift the leaf out of the ethanol, dip it into the hot water to soften it, spread it out on a white tile and cover it with iodine solution. A blue-black colour indicates the presence of starch in the leaf.

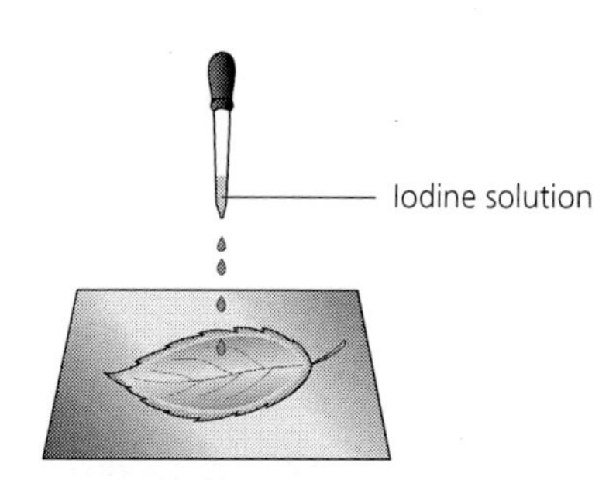

A variegated pelargonium leaf

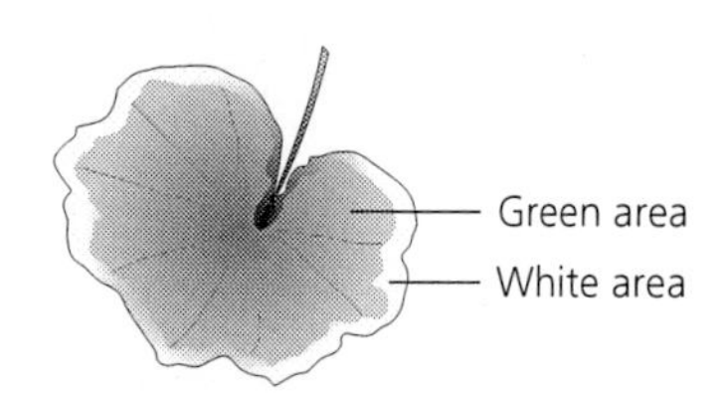

Your teacher will be looking for:

- careful and safe use of the apparatus given
- accurate observations
- good presentation of results
- sensible conclusions which fit your results

HAZARD WARNING

Ethanol is highly flammable.
KEEP AWAY from naked flame.
WEAR EYE PROTECTION.

?

You are provided with variegated leaves – leaves which have areas with and without chlorophyll. Design an experiment using these leaves to show that chlorophyll is necessary for photosynthesis.

 Name:

Light, carbon dioxide, and photosynthesis

You need:

- potted plants
- scissors
- ethanol
- white tiles
- boiling tubes
- 500 cm^3 beakers
- Bunsen burner, tripods and gauzes
- black paper or polythene
- iodine solution
- paper clips
- soda lime
- conical flasks
- clamps and stands
- cotton wool
- vaseline
- spatulas

In this exercise, you will try to work out how photosynthesis is affected by differing levels of light and carbon dioxide.
The practical method has not been devised for you.

Do plants need light for photosynthesis?

1. You could test two plants for starch – one that had been in the dark for 12 hours and one that had been in the light. Can you think of more interesting experiments using the materials provided?
2. How could you use strips of black paper or polythene, or even a black and white 35 mm photographic negative?
3. At which stage will you detach the experimental leaf from the plant?
4. Predict what a leaf will look like after the starch test. Make drawings of your results.

Do plants need carbon dioxide for photosynthesis?

Extension

- You could use the apparatus in the drawing opposite.
- Why must you start with a plant which has been in the dark for 12 hours?
- Why must you test a leaf for starch before setting up the apparatus opposite?
- Find out what the soda lime will do to air in the flask.
- Why is the vaseline necessary?
- What control is needed?
- Where will you put the plant and for how long?
- What test is needed to obtain a result?
- How will you present results and conclusions?

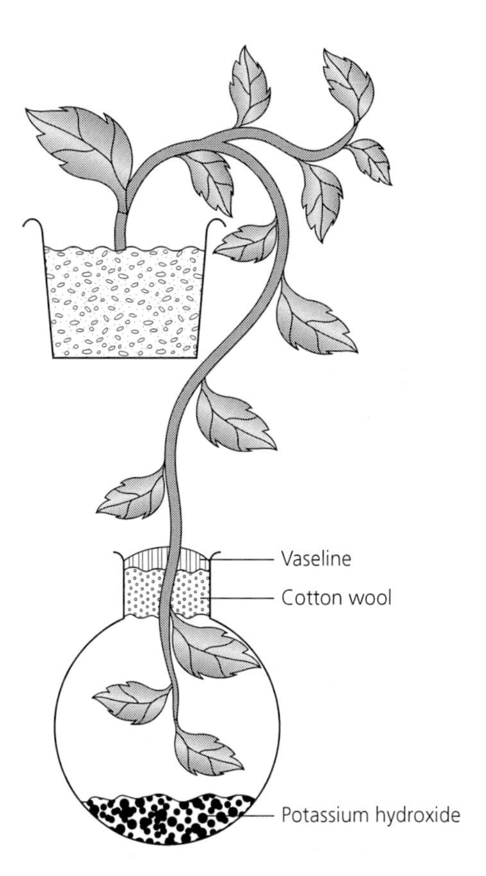

Your teacher will be looking for:

- good experimental planning
- careful and safe use of the apparatus given
- good observation and presentation of results
- sensible predictions (hypotheses) and conclusions

HAZARD WARNING

Ethanol is highly flammable. KEEP AWAY from naked flame. Soda lime (sodium hydroxide and calcium hydroxide) is CORROSIVE and can cause severe burns; also dangerous to eyes and skin. AVOID SKIN CONTACT. AVOID CONTACT WITH WATER. WEAR EYE PROTECTION.

 Name:

Evolution by natural selection

You need:

- sheet of white paper
- newspaper
- forceps
- coloured pencils
- clock/watch with 'second' hand
- 30 newspaper circles (made with hole punch)
- 30 white circles (made with hole punch)

In this exercise, you will simulate how predators locate prey in different environments. You will analyse how camouflage (colour and pattern) affect an organism's ability to survive in certain environments.

Method

1 Work in pairs. Place a sheet of white paper on the table. One student should spread 30 white circles and 30 newspaper circles over the surface while the other student (the 'predator') isn't looking.

2 The 'predator' will then use forceps to pick up as many of the circles as he or she can in 15 seconds. This corresponds to the predator capturing and eating the prey species.

3 This trial should be repeated with white circles on a newspaper background, newspaper circles on a white background, and newspaper circles on a newspaper background.

Record the data in the chart below.

		Starting Population		Number 'eaten'	
Trial	Background	Newspaper	White	White	Newspaper
1	white	30	30		
2	white	30	30		
3	newspaper	30	30		
4	newspaper	30	30		

Questions

a What type of tree surface is represented by the 'newspaper' background?

b Which moth coloration (pale or dark) is the best adaptation for a 'smoke-polluted' background? How do you know?

c Following trial 1, what has happened to the frequency of the allele for 'light' colouration?

d Moths which survive i.e. are not eaten by predators, can pass on their alleles when they reproduce. How does the simulation model natural selection?

Extension

- Find out the meaning of the term **Industrial Melanism**.
- Hospital managers are very worried about hospital infections. Explain how natural selection might lead to antibiotic-resistant strains of bacteria.
- How do the stripes on a zebra help it to avoid predators? Do the stripes on a tiger have the same effect?

PRACTICAL **Name:**

Transport tissue in plants

You need:

- microscopes
- slides and coverslips
- razors
- celery
- germinated broad beans
- white tiles
- Petri dishes
- paint brushes
- eosin dye

Water-conducting tissue of celery

1 Obtain a stick of celery, preferably with leaves still attached. Put it in a beaker half-filled with eosin dye and leave it for 24 hours.

2 Look carefully at the leaf veins. Observe and describe what has happened. Explain what has happened.

3 Lay the celery on a white tile and use a razor to cut this slices off it. Continue until you have a slice so thin it is almost transparent.

4 Use a paint brush to transfer the slice to a microscope slide, add a drop of water and lower a coverslip over it.

5 Make a drawing of the slide showing which areas have turned red. What are these areas? Refer to page 66 of the text book.

Celery
Razor blade
White tile

Compare root and stem of beans

1 Germinate a number of broad bean seeds by trapping them against the sides of a jam jar with a cylinder of blotting paper filled with damp sand or sawdust. Leave them until the root and stem have developed.

2 Clamp a bean over a beaker of eosin so that its root is immersed in the dye. Leave it until the dye becomes visible in the leaf veins.

3 Cut thin slices of the root and stem, and make drawings to show which areas have been stained red.

What is the difference between the position of xylem in a bean stem and root?

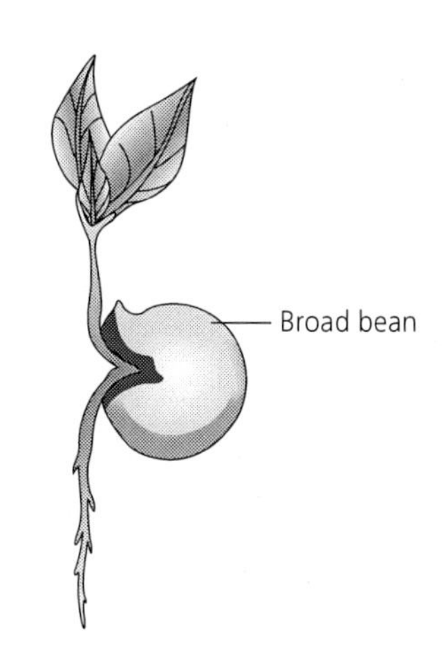

Your teacher will be looking for:

- careful use of the apparatus given
- good observation
- good presentation of results as diagrams

HAZARD WARNING

Razor blades are sharp, handle with care.

PRACTICAL **Name:**

Measuring transpiration

You need:

- privet or other leafy twigs
- capillary tubes
- vaseline
- cotton wool
- rubber tubing
- labelled specimen tubes

Making a potometer

1 Set up the apparatus opposite in the following way. Push the piece of rubber tubing about 2 cm over the end of the capillary tube, submerge in water and squeeze the rubber tube until *both* tubes are full. Push the end of the twig into the rubber tube (don't wet the leaves) and, without letting water escape, clamp the apparatus over a beaker of water. Seal all joints with vaseline and fasten a card scale in position (see diagram).

2 After five minutes raise the apparatus so the capillary is out of the water. Air should start to move up the capillary. How far does it move in two minutes? Find the average time for three, two-minute runs.

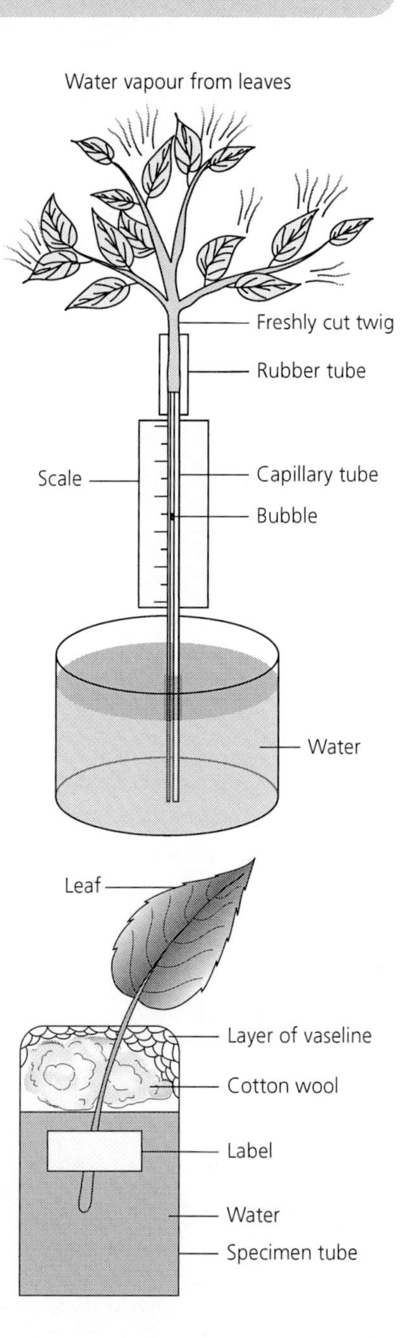

Questions

a What *exactly* have you measured?

b Is what you measured the same as transpiration?

c How could you measure the *volume* of water the twig takes up each minute?

The effects of climate on transpiration

Method one

Devise an experiment using the apparatus above to discover the effects on the twig of the five *different* climatic conditions listed in method two.

Method two

What *measurable* change will take place in the apparatus opposite with time? What will cause this change?

Design an experiment using this apparatus to discover the effects on transpiration of hot, cold, windy, humid and dark conditions.

Start by thinking about the following:

- How you will create these conditions?
- What will you measure, and when?
- What controls are needed and how will you record your results? Remember: one condition is changed; others remain the same.

Your teacher will be looking for:

- careful use of the apparatus given
- accurate observation and measurement
- good presentation of results
- good experimental design and use of controls
- sensible conclusions

Name: ....

Stomata

You need:

- microscopes
- slides and coverslips
- fine-point scissors
- sugar
- long balloons
- Visking tubing
- cotton thread
- transparent sticky tape
- privet, or a locally available evergreen species

Looking at stomata

1 It is possible to tear privet (or evergreen) leaf so that a small area of lower skin is exposed. Hold a leaf with its lower surface uppermost and tear it diagonally rather than straight across.

2 Make a microscope slide of lower skin and look at it under a magnification of at least × 600.

3 Estimate the size, in micrometres, of stomata. Roughly how many stomata are there per square millimetre of leaf surface?

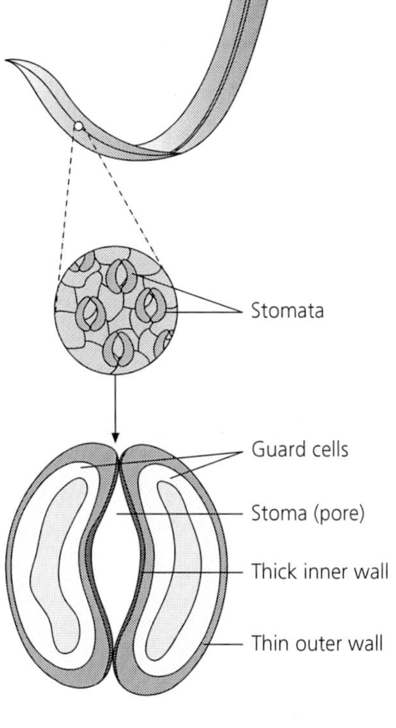

A balloon model of a guard cell

1 Fasten a length of sticky tape to one side of a long balloon. Inflate the balloon and compare its shape with another balloon without the sticky tape.

2 Guard cells have thicker walls next to the stoma than on their outer walls (diagram above). Think about the shape of the balloon with the sticky tape, and formulate an hypothesis to explain the function of a guard cell's thick inner wall. What effect could it have as guard cells inflate and deflate to open and close a stoma?

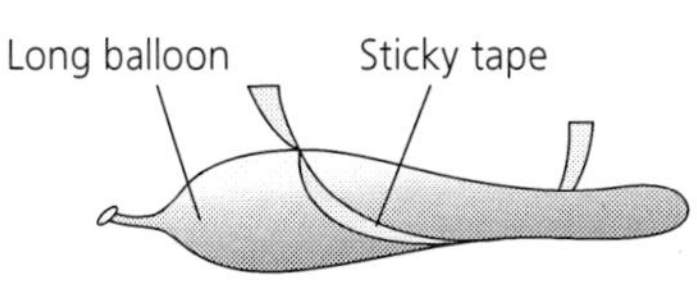

More things to do

Make a Visking tube model of stomata.
Fill two 20 cm lengths of Visking tube with strong sugar solution and seal both ends with cotton. Tie the two Visking sausages firmly together, place them in water for 30 minutes and explain what happens.

What do the two Visking sausages represent?
Use this result to formulate an hypothesis to explain how stomata open and close.

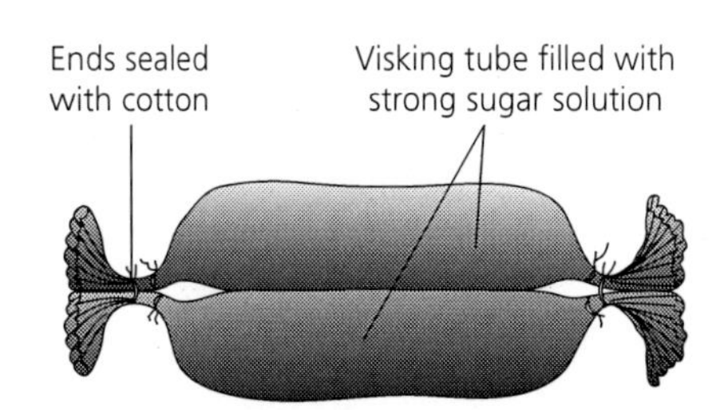

Your teacher will be looking for:

- careful use of the apparatus given
- good observation of the privet leaf stomata under the microscope
- good understanding of the behaviour of stomata as shown by models

 Name:

Demonstrating respiration

You need:

- boiling tube and rubber bung with two holes
- limewater
- glass tubing
- four conical flasks and four rubber bungs with two holes each
- rubber tubing

Compare breathed and unbreathed air

1 Prepare the apparatus shown below, on the left. How long does it take for limewater to turn milky when you blow air gently down tube A? Wash out the boiling tube and refill with fresh limewater. How long does it take for limewater to turn milky when you suck air gently through tube B?

Questions

a What turns limewater milky?

b What do your results tell you about the difference between laboratory air and breathed air?

c What exactly do your results prove?

Demonstrate respiration in animals and plants

1 Set up the apparatus on the right below. What is the purpose of flask 1? Why is flask 2 necessary? What does the apparatus demonstrate?

2 Investigate woodlice, maggots, earthworms, etc., in a small specimen chamber. Use a bell jar to investigate bigger animals.

Questions

a How could you use this apparatus to make rough comparisons of respiration rate in, for example, woodlice and earthworms?

b What things would you have to keep the same?

c How would you set up the apparatus to demonstrate respiration in plants?

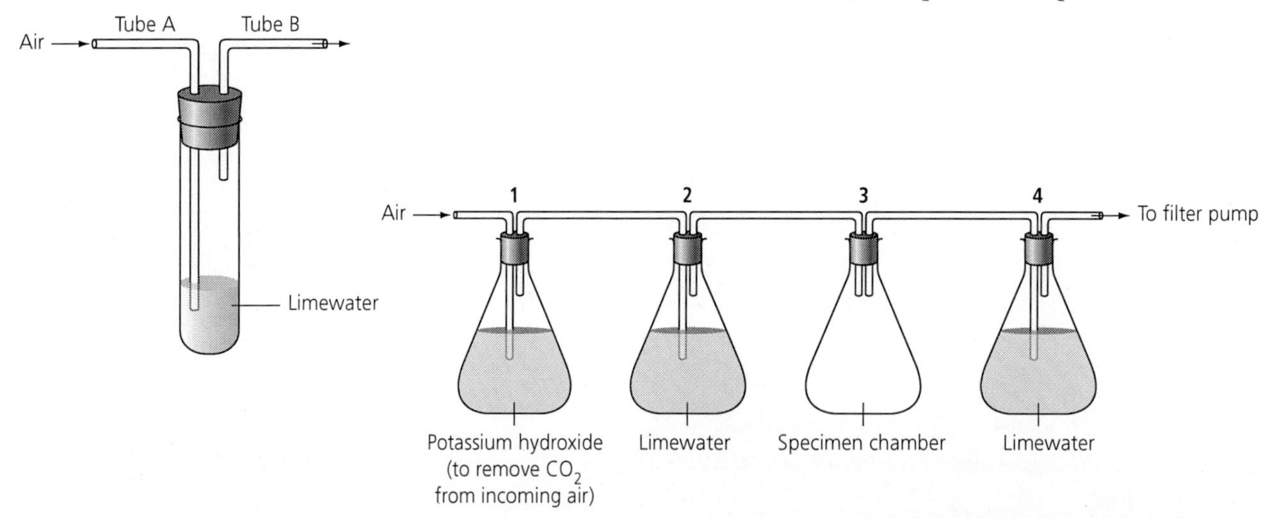

Your teacher will be looking for:

- accurate observation
- good presentation of results
- care with live specimens
- sensible conclusions which match your results

HAZARD WARNING

Soda lime is CORROSIVE and can cause severe burns; Also dangerous to eyes and skin.
AVOID SKIN CONTACT.
AVOID CONTACT WITH WATER.
WEAR EYE PROTECTION.

Name: ....

Pulse and breathing rates

You need:

- clock or watch with a 'second' hand
- bench or step about 30 cm high

A standardised fitness test

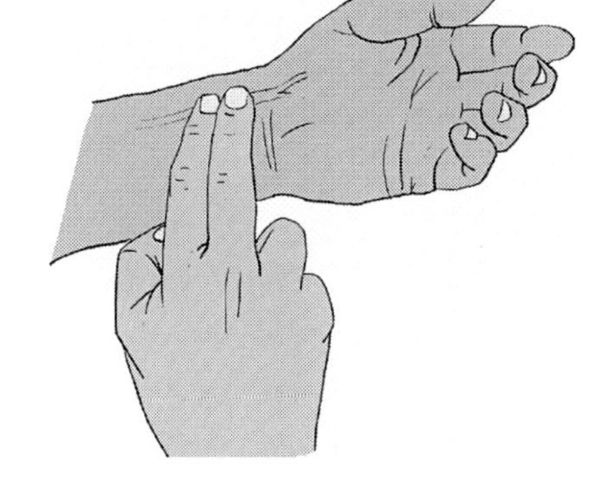

(Check with your teacher that you should do these exercises *before* beginning this activity.)

1 Form groups of three. You will take it in turns to do a standardised exercise. (The two members of the group who are not exercising keep time and measure breathing and pulse rates.)

2 Before you start, measure and record your resting pulse rate and your resting breathing rate using the methods below:

> **Pulse rate:** Place your fingers as shown in the diagram. Move them until you can feel a pulse. Count the pulses in 30 seconds, then multiply by two to get the pulse rate per minute.
>
> **Breathing rate:** Count the number of breaths taken in 30 seconds and multiply by two.

3 One member performs the following exercise while another calls out the time at one minute intervals.

4 Face the bench. Step onto it with one foot, then step up with the other foot. Step down with the first foot, then down with the second foot. Practise this so that your can step steadily at about 25 steps per minute.

5 When you and the timer are ready, start stepping and keep going at a steady speed for three minutes.

6 *Immediately* afterwards, one person measures the pulse rate of the 'exerciser'. At the same time, the other team member measures the exerciser's breathing rate. Keep a record of your results.

7 Continue monitoring the pulse rate and breathing rate at one minute intervals until they return to normal.

8 Swap places and repeat steps 2 to 6 until you have all done the test.

HEALTH CHECK

Check with your teacher that you are able to participate in this activity.

Questions

a What are the fastest, slowest, and average *resting* pulse rates for the class?

b What are the fastest, slowest, and average *resting* breathing rates for the class?

c What are the fastest, slowest, and average pulse rates *after exercise* for the class?

d What are the fastest, slowest, and average breathing rates *after exercise* for the class?

e Is the fittest person the one with the lowest or the highest pulse rates?

f How does fitness affect the time taken for the pulse to return to normal?

PRACTICAL **Name:**

Measuring your fitness level

You need:

- stool, step or wooden box; about 30 cm high
- stopclock or watch (with 'second' hand)

Method

1 Step up and down at a rate of five times per ten seconds, for five minutes.

2 Rest for one minute: after 15 seconds of this rest your partner should measure and record your pulse count for 30 seconds. The easiest pulse to find is the temporal pulse – just where the top of your ear joins the side of your head.

3 Repeat this process for a second minute, and then for a third minute. Your partner should by now have three recordings of your pulse count – each of these is the number of heartbeats in 30 seconds.

4 Calculate your fitness index using this calculation:

Time of exercise in seconds (X)

Mean value for pulse count (Y)

Fitness index = $X/Y \times 10$

(a value above 7.5 means you are very fit, 6.0 means you are quite fit, below 5.0 means you are not very fit)

Questions

a Apart from the change in pulse rate, suggest two other changes that take place during periods of exercise. What is the value of these changes to the body?

b Why do you think a professional footballer is likely to have a higher fitness index than a sprinter?

Extension

What are the properties of the heart and the arteries that give you a pulse?

PRACTICAL **Name:** ....

Strength and muscle fatigue

You need:
- large textbooks or other heavy objects
- stop watches or clocks with 'second' hands

A test of strength

1 Work in pairs. One student makes accurate timings, in seconds, while the other performs a task.

2 One student must hold a book or other heavy object in his or her preferred hand (i.e. right hand if right-handed) so that it is at arm's length. The arm must be straight out at the shoulder and must be kept in this position for as long as possible.

3 The time-keeper starts timing as soon as the book is in the required position and stops when the book is lowered or the arm bent. Record how long the weight is held.

4 The pair now swap tasks and repeat this procedure.

The pair swap tasks again but this time the book is held in the *non-preferred hand.*

The pair swap tasks again so that the other student tries the task with his/her non-preferred hand.

5 Results for each student should be recorded on the chalkboard, then converted into a table showing how many students held the weight, in each hand, for intervals such as 60–69 seconds, 70–79 seconds, etc. up to the longest time recorded.

6 Convert these results into histograms for preferred and non-preferred hands. Explain any differences between preferred and non-preferred hands.

Explain any differences between the sexes. What caused fatigue in muscles?

Investigating muscle fatigue

7 Use the results of the previous experiment to find four students who are roughly equal in strength.

8 **Student one** holds a weight at arm's length, as described in the previous experiment, for as long as possible. He/She rests for 10 seconds and then repeats the task.

Continue in this way for five repetitions or until the student is too tired to go on. Record, in seconds, the time the weight is held for each repetition.

9 **Student two** does the same as student one except that a rest of 20 seconds is allowed between repetitions.

10 **Student three** does the same as student one except that a rest of 30 seconds is allowed between repetitions.

11 **Student four** does the same as student one except that a rest of 40 seconds is allowed between repetitions.

12 Construct a table showing how long each student held the weight during each repetition.

Questions

a What effect on performance did the different rest periods have?

b Try to explain these differences (relate them to the efficiency of the circulatory system, anaerobic respiration and the oxygen debt).

Your teacher will be looking for:
- the use of 'fair' tests to measure strength and muscle fatigue
- good presentation of results in tables and as histograms
- sensible conclusions consistent with your results

HEALTH CHECK

Check with your teacher that you are able to participate in this activity.

 Name: ....

Measuring respiration

You need:

- specimen tubes and bungs with two holes
- soda lime or potassium hydroxide (0.4% concentration)
- 50 cm^3 beakers
- capillary tubing
- perforated zinc
- ink (coloured water)
- white card
- sticky tape

Method

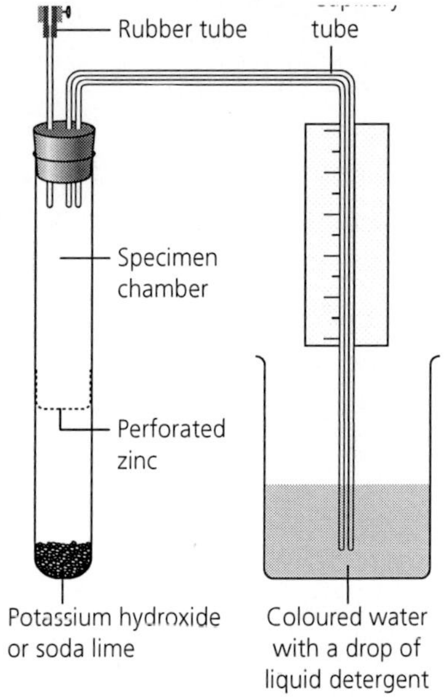

1 Prepare the apparatus opposite. It is a simple **respirometer**; that is, it can be used to measure the rate of respiration. Predict what will happen to coloured liquid in the capillary tube if you put a few maggots or other small creatures in the specimen chamber and close the screw clip. What is the reasoning behind your prediction?

2 Carry out this experiment and check your prediction. Aerobic organisms take in oxygen and produce carbon dioxide at about the same rate. Carbon dioxide is absorbed by the soda lime. Use these facts to explain your result.

Rubber tube
tube
Specimen chamber
Perforated zinc
Potassium hydroxide or soda lime
Coloured water with a drop of liquid detergent

How to measure respiration rate in a variety of small animals

What exactly does this apparatus measure?

Design an experiment to measure and compare respiration rate in small creatures such as woodlice, beetles, spiders, earthworms, maggots, etc.

Start by thinking about the following:

- What conditions must be kept the same for each creature so that results can be compared?
- How could you improve your results if you knew the diameter (bore) of the capillary tube?
- Why is the rubber tube and screw clip essential, and when should the clip be opened and closed?

Your teacher will be looking for:

- careful use of the apparatus given
- suitable treatment of living specimens
- accurate observation and measurement
- good presentation of results
- sensible conclusions

HAZARD WARNING

Soda lime is CORROSIVE and can cause severe burns; also dangerous to eyes and skin. AVOID SKIN CONTACT. AVOID CONTACT WITH WATER. WEAR EYE PROTECTION.

Name: ....

A model for breathing movements

You need:

- large plastic syringe
- cork/rubber bung
- balloon

The action of the diaphragm

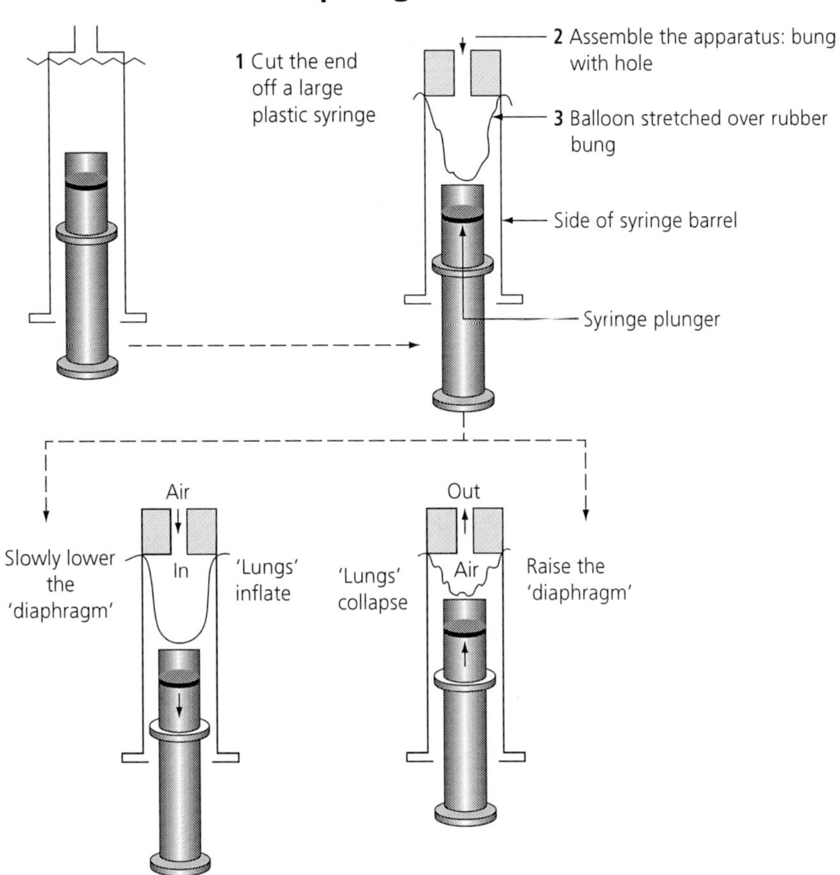

Questions

a What does the hole in the rubber bung represent?

b What does the balloon represent?

c What does the end of the plunger represent?

d The rigid side walls of the syringe barrel represent the wall of the thorax. What makes up the wall of the thorax in a living animal?

Extension

In the body of a mammal certain structures pass through the diaphragm. Name any **three** of them, and explain why they must pass through the diaphragm.

PRACTICAL **Name:**

The ribs and intercostal muscles

You need:

- wooden or plastic strips, with holes drilled in them (items from basic construction kits are ideal)
- nuts and bolts
- elastic bands
- matchsticks

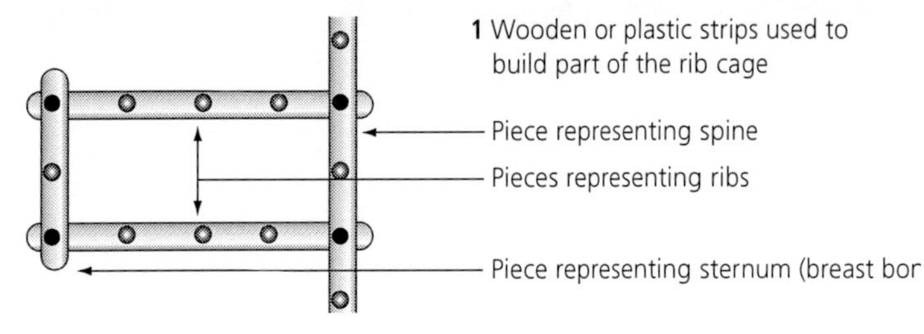

The demonstration needs two students to work in a pair
The first student holds the 'breast bone' and 'spine' firmly
The second student fits the 'external intercostals muscles' between the ribs.

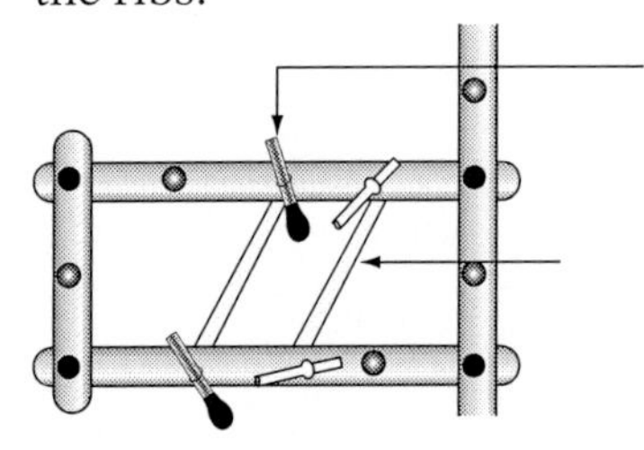

2 Matchsticks and elastic bands are used to represent the intercostals muscles and their attachments to the ribs

The first student allows 'breast bone' and 'ribs' to move when the 'external intercostal muscles' contract

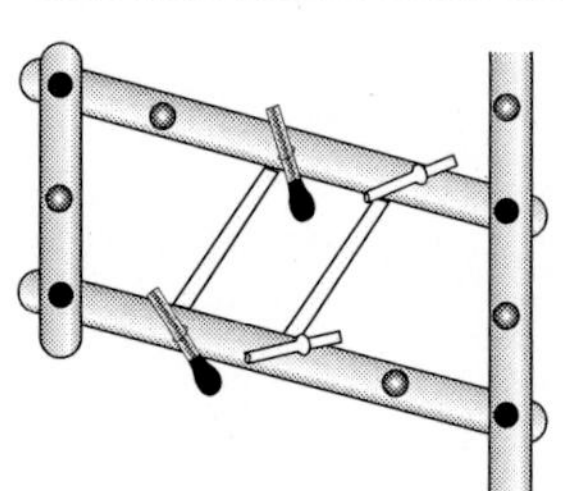

3 Ribs move upwards and outw

Questions

a Explain how these movements of the ribs cause air to be inhaled.

b As well as the *external* internal muscles, there is a set of *internal* intercostals muscles. What is the function of this set of muscles?

Extension

What type of tissue would you expect to find where the ribs meet the spine, and where the ribs meet the sternum? Explain why it is important that this tissue is present.

Sometime the nervous control of breathing fails (perhaps because the brain control centre is damaged). If this happens, an artificial ventilator may be needed. Explain how this machine works.

PRACTICAL **Name:** ....

Anaerobic respiration

You need:

- test tubes
- thermometer
- balloons
- measuring cylinder
- limewater
- liquid paraffin
- Thermos flasks
- yeast in cold, boiled water
- glucose in cold, boiled water
- fresh peas in cold, boiled water
- boiled peas in cold, boiled water containing bactericide (eg. sodium chlorate(I) 10%)

Anaerobic respiration (fermentation) in yeast

1 Label three test tubes A, B, and C.
Place 20 cm^3 of yeast suspended in cold, boiled water into tube A. (Boiling removes oxygen from water.) Add a few drops of liquid paraffin (to cover the surface of the yeast suspension). Place 20 cm^3 of glucose dissolved in cold, boiled water into tube B. Add liquid paraffin. Place 10 cm^3 of yeast suspended in cold, boiled water into tube C, then add 10 cm^3 of glucose dissolved in cold, boiled water. Mix the two together. Add a few drops of liquid paraffin.

2 Place a balloon firmly over the neck of each tube (tie with cotton if necessary). Make sure the balloon is deflated. Put the tubes in a warm place for 24 hours.

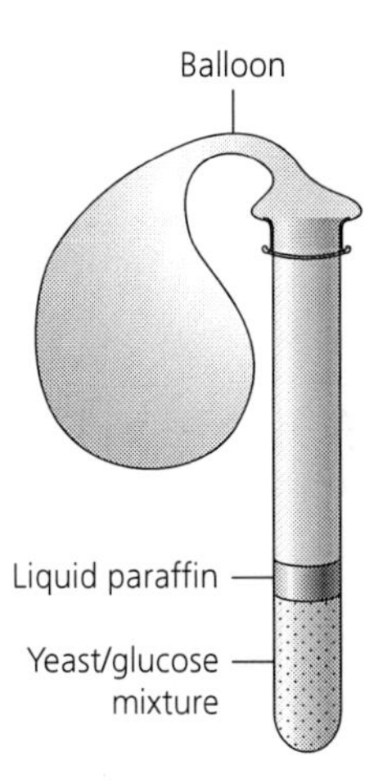

3 Record what happens to the balloons.
Explain your observations.
Why was cold, boiled water used in this experiment?
Why was liquid paraffin added to each tube?
What gas entered a balloon?

Anaerobic respiration in peas

1 Fit two Thermos flasks with bungs through which a thermometer has been passed. Label the flasks A and B.

2 Fill flask A with boiled peas in cold, boiled water containing bactericide.
Fill flask B with fresh peas in cold, boiled water.
Note the temperature of each flask.

3 Place the bung in each flask, so that air cannot enter and leave them for a week.
Note any temperature changes daily.

4 Explain any changes which occur in the temperature, and in the peas.
Why was one set of peas boiled?
Why were both sets of peas in boiled, cold water.
Why was bactericide added to flask A?

Your teacher will be looking for:

- careful and skilful use of the apparatus given
- good observation and presentation of results
- sensible conclusion which match your results

HAZARD WARNING

Sodium chlorate(I) solutions is corrosive. AVOID SKIN CONTACT. WEAR EYE PROTECTION.

PRACTICAL Name:

Lung volume

You need:

- bell jar with 500 cm^3 graduations down one side
- 50% alcohol
- rubber tube
- deep sink

Breathe in as such air as you can, then breathe out as much air as you can. The volume of air which you breathed out is called the **vital capacity** of your lungs.

After you have forced as much air as possible out of your lungs there is still about 1500 cm^3 left behind. This is called the **residual volume** of your lungs.

The **total volume** of your lungs is your vital capacity *plus* your residual volume.

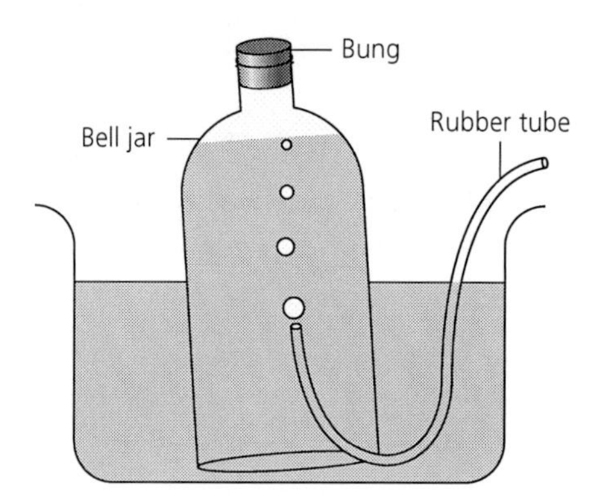

Method

1 Mark out the side of a bell jar in 500 cm^3 units as follows:
Place a bung firmly in a bell jar and turn it upside down in a sink (bung downwards). Pour 500 cm^3 of water into the bell jar and mark the water level with a chinagraph pencil or indelible marker.

Repeat until the whole of one side is marked out.

2 Fill a deep sink with water. Place the bell jar on its side in the water so that it is completely full. Without letting in any air, turn the bell jar upright in the sink. Run water out of the sink until it is about half full.

3 Work in pairs.
One student tilts the bell jar enough to let in a rubber tube (see diagram) and then holds the tube in place.

Sterilise the end of the rubber tube in alcohol, then wash it in water. The other student breathes in as much air as possible and then blows out as much air as possible through the rubber tube into the bell jar.

Raise the bell jar so that the level of water inside is level with water in the sink, and read off and record the vital capacity of your lungs.

Sterilise the end of the rubber tube in alcohol, then wash it in water.
Swap tasks and repeat this procedure. Prepare a class results table on the blackboard.

Questions

a What is the total volume of your lungs?

b What is the average vital capacity of your class?

c What is the smallest and the largest vital capacity in your class?

Your teacher will be looking for:

- accurate measurement and recording of results
- good presentation of class results and accurate calculation

HAZARD WARNING

Alcohol is highly flammable.
KEEP AWAY from naked flame.
WEAR EYE PROTECTION.

PRACTICAL **Name:** ....

The functions of perspiration

You need:

- thermometers
- cotton wool
- small beakers of water and alcohol

Method

1 Wave your hand backwards and forwards in front of you. Does it feel warmer or cooler?

2 Swab a little water onto the back of one hand with cotton wool and wave it backwards and forwards again. What does the wet part feel like compared with the dry parts?

Note any feelings to do with temperature sense.

3 Dry your hand, then swab a little alcohol onto it with cotton wool. Wave it backwards and forwards again. What does the alcohol-treated part feel like compared with the dry parts?

Note any feelings to do with temperature sense.

Does the alcohol give you a sensation which is different from water?

4 What conclusion can you draw at this stage of the experiment?

5 Obtain three thermometers. Record the temperature of each. Wrap the bulbs of all three thermometers in a thin layer of cotton wool. Tie it in place with cotton. Leave the thermometers to acclimatise for five minutes, then record the temperature reading of each.

6 Take one thermometer (with dry cotton wool) and wave it for one minute, then record the temperature reading. Take another thermometers, dip it into water, wave it for one minute, then record the temperature reading. Take a third thermometer, dip it in alcohol, wave it for one minute, then record the temperature reading.

Questions

a How do the results with a thermometer dipped in water and alcohol compare with those from the thermometer covered with dry cotton wool?

b How do the results of the thermometer experiment help explain the results from treating skin with water and alcohol?

c Alcohol evaporates more quickly than water. Use this information to explain results from both experiments. What do these results tell you about the function of perspiration?

?

A chemist has invented a new, harmless liquid that evaporates quickly. She wants to sell it in swabs which people can use to wipe themselves when they are too hot.

Design an experiment to compare the effectiveness of this invention with a wet cloth.

Your teacher will be looking for:

- careful and safe use of the apparatus given
- accurate observation
- good presentation of results
- sensible conclusions

HAZARD WARNING

Alcohol is highly flammable.
KEEP AWAY from naked flame.
WEAR EYE PROTECTION.

PRACTICAL **Name:** ....

Heat loss from a model body

It is possible to represent the body of a human by a model made up of a boiling tube two-thirds filled with water. If the temperature of the water is raised, and a temperature gradient is established between the tube contents and the surroundings, it is possible to investigate heat loss from the body. To ensure measurable changes, this investigation requires that you pre-heat the contents of the tube to boiling point and construct cooling curves to demonstrate heat loss over a fifteen minute period.

You need:

- 3 boiling tubes
- −10 to +110 °C thermometer
- 1 dm^3 beaker
- 2 retort stands with clamps
- Bunsen burner, tripod, gauze, heat resistant mat
- test tube holder
- 2 squares of denim cloth, each large enough to wrap around the boiling tube to give a single layer of material
- 2 squares of aluminium foil, each a little (1 cm all round) larger than the pieces of denim
- stopwatch
- test tube rack
- wash bottle
- hair drier

Method

1. Assemble the apparatus as described in the introductory paragraph, and construct a cooling curve for the uncovered boiling tube. Now use the apparatus to investigate the statement that:

   "Denim jeans are adequate for a low level stroll, but wet denim on a windy hillside represents a potentially lethal combination"

2. Present your results in a suitable table. Plot a graph of your results.
3. Comment on the significance of your results.
4. List any major sources of error, and suggest improvements which might be made to the experimental technique.
5. Suggest extensions of the investigation which you might make, using the same or very similar apparatus.

 Name: ....

Skin and sensitivity

You need:

- pair of dividers (a metal hairpin will do)
- mounted needle (pin will do)
- ruler with millimetre scale

Since the skin is in direct contact with the environment, you should not be surprised that it has many sensory cells within it. In this series of experiments you will be able to investigate the distribution and the method of action of some sense cells in the skin.

Method

Work in pairs throughout these experiments.

Variation in sensitivity of different regions of the skin

1 Use a pair of dividers to apply two simultaneous touch stimuli to the outside of your partner's forearm. The points should be exactly 4 cm apart and the subject (blindfolded or looking in the opposite direction) should say when he or she feels them as two separate sensations.

2 Now reduce the distance between the points of the dividers and determine the minimum distance by which the stimuli must be separated for the subject to feel both of them (in other words, find the distance at which the two points seem to be causing a single stimulus).

3 Repeat the experiment on other parts of the body as listed below. Take great care when working on or near the subject's face. In each case record the distance at which the two stimuli are 'sensed' as one.

   Outside of forearm; fingertips; back of hand; palm of hand; lips; shin; back of neck; inside of forearm; sole of foot.

4 Record all of your results in a suitable table.

   Calculate the mean values for your class, and present them in the form of a table.

5 Now explain your results – why are some parts of the skin more sensitive than others (use a simple diagram to help your explanation)

   What is the relevance of your results to blind people who read Braille?

Adaptation to stimuli

1 With a mounted needle wiggle one of the hairs on your partner's arm or hand until he or she can no longer feel it. How long did it take for the sense cell at the base of the hair to adapt?

   Why is it an advantage to get used to stimuli of this kind?

2 Wiggle another hair just long enough for your partner to appreciate the sensation. Now rub the skin vigorously with your finger for 15 seconds, and wiggle the hair again. Can he or she still feel it? Suggest an explanation for what has happened.

PRACTICAL **Name:** ....

Measuring the speed of reflexes

You need:

- metric rulers

Measure the speed of your reflexes

1 Work in pairs. One student holds a ruler between thumb and forefinger so that the ruler hangs with its zero mark at the bottom. The other waits with thumb and forefinger of one hand about 2 cm apart and level with the zero mark of the ruler.

2 The student holding the ruler says 'ready', then drops the ruler within five seconds without further warning. The other student must catch the ruler between thumb and forefinger.
Note the number of centimeters the ruler has dropped by looking at the position of the thumb and forefinger on the ruler.

3 Calculate the average distance over at least ten ruler drops. Use the graph opposite to convert this distance into response time, in seconds. Draw a graph showing the range of results for the whole class.

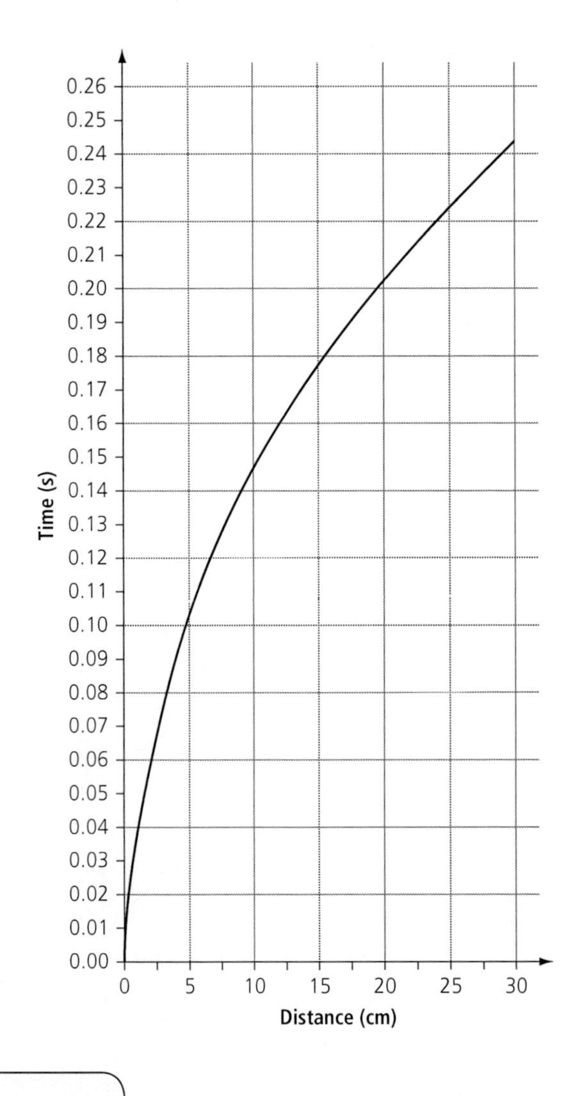

Questions

a Name all the parts of the nervous system which impulses travel through as you respond to the ruler dropping.

b Your result is the time it takes for impulses to travel from your eyes to your hand. Measure this distance and use it to calculate the speed of nerve impulses, in meters per second.

Your teacher will be looking for:

- careful collection of data
- good presentation of results
- accurate use of the conversion graph
- successful calculation of the approximate speed of nerve impulses

Further work

The 'detector' in the experiment above was the eye. Design an experiment to measure the speed of reaction when the signal is a sound detected by the ear.

PRACTICAL | Name:

Bones

You need:

- chicken leg bones
- safety goggles
- retort stands and clamps
- dilute hydrochloric acid
- masses and mass hangers
- Bunsen burner
- metre rule
- tongs

Investigating bone structure

Bone contains **minerals** (mainly calcium compounds) and **organic fibres.** What is the function of these parts?

1. Obtain two small bones (e.g. chicken leg bones). Remove the minerals from one bone by soaking it in dilute hydrochloric acid for 24 hours. Soak another bone in water for the same time (the control).
2. Pour away the acid, wash the bone and try to bend it. Compare it with the bone soaked in water. What do your observations tell you about the functions of minerals and organic fibres in bones?
3. Remove the organic fibres from another small bone by holding it (with tongs) in a hot Bunsen burner flame for a few minutes in a fume cupboard.
4. When it is cool, try to crush the burnt end with a pencil or stick. What do your observations tell you about the functions or organic fibres in bones?

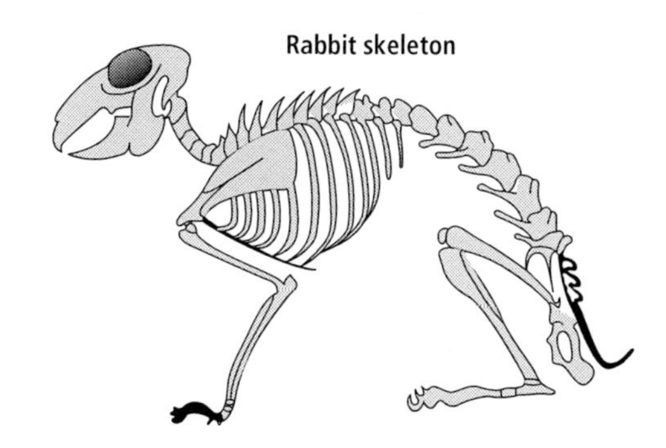

Why are backbones arched?

Look at the skeleton of a rabbit or other four-legged animal and note the shape of its backbone. Is there an advantage in having an arched backbone?

1. Support a metre rule (or strip of wood) on chair backs or clamp stands so that each end overlaps the support by 1 cm. Attach weights to its centre until it bends and falls off the supports.
2. Place a metre rule between two supports in such a way that when the supports are moved towards each other the rule arches upwards slightly.
3. Attach weights to its centre as before. What do your results tell you about the advantage of having an arched backbone?

Your teacher will be looking for:

- careful and safe use of the apparatus given
- accurate observation
- good presentation of results
- sensible conclusions

HAZARD WARNING

Hydrochloric acid is harmful. Avoid skin contact. WEAR EYE PROTECTION. When breaking the burnt bones WEAR EYE PROTECTION and use a safety screen.

PRACTICAL **Name:**

Antagonistic muscles at the elbow

You need:

- stiff cardboard
- scissors
- needle and thread
- paper fastener

Method

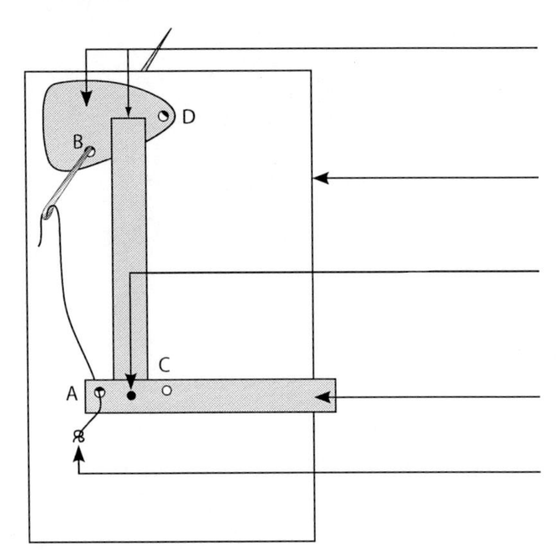

1 Draw in the shoulder girdle (scapula) and the humerus, and bore the holes B and D.

Backing card

2 Cut the strip of cardboard to represent the ulna and radius, bore the holes A and C, and use a paper fastener to loosely attach the ulna-radius to the humerus.

3 Thread a strand to represent the biceps muscle, and another to represent the triceps muscle.

4 Pull on the strands, one at a time. What is the effect of pulling on the triceps?

Questions

a The biceps and triceps are **antagonistic muscles.** What does this mean?

b How is an instruction to 'contract' given to a muscle?

c Muscles are well supplied with blood. Name two things that the blood supply must deliver to the working muscle, and two things that must be transported away.

Extension

Draw lines on your diagram to represent the two muscles which could move the whole arm backwards and forwards (as in a running movement, for example).

Your model represents the ulna and radius as one 'piece'. In fact the ulan and radius are two separate bones. Why is it an advantage to have two bones in the human forearm?

PRACTICAL **Name:**

Muscles and tendons

You need:

- leg from a chicken
- forceps
- scalpel
- scissors
- mounted needle

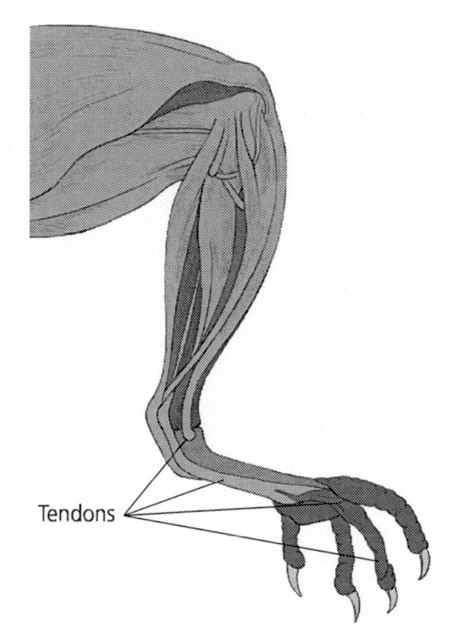

Method

1 Cut the skin away from the leg of the chicken so that you can see the muscles and the tendons.

2 Use the mounted needle and scalpel to separate the muscles and tendons from one another.

3 Pull each of the tendons in turn. Record what happens when you do this. As a result of what you see you may need to separate the muscles and tendons more carefully.

Tendon	A	B	C	D	E	F
What happened						

Tendon	G	H	I	J	K
What happened					

Questions

a What is the purpose of the tendons in a living bird?

b Tendons contain many fibres of a protein called **collagen**. This means that they will not stretch. Why is this important?

c Each of these tendons is attached to a muscle. Use your observations to explain what is meant by the word '**antagonistic**' when it is used to describe the action of muscles.

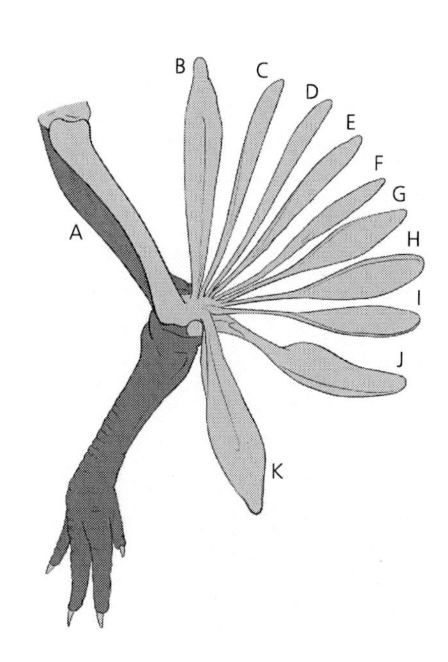

Extension

Where is your Achilles tendon? Which muscle is it attached to? What happens when this muscle contracts?

Try to find out why this is called the **Achilles** tendon.

PRACTICAL Name:

Eyes and vision

You need:

- two pencils
- large coins
- flat desk or table top

Distance Judgement

This experiment shows you how to compare one eye with two eyes when judging distances. It should help you understand why we have two eyes.

1 Arrange two pencils on a desk top in positions A and B, as shown on the diagram opposite.

2 Sit so that your eyes are level with the surface of the desk (it is very important that you do not look down on the desk).

3 The aim is to move the pencils until their points are exactly opposite by *not touching* (i.e. to positions C and D). Do this in three different ways:

Method one: Close one eye and ask a partner to move the pencils by *following your instructions only* (they must not try to correct your mistakes).

Return the pencils to positions A and B.

Method two: Close one eye but this time use *one hand* to move the pencils.

Method three: Try the experiment again using one hand but *both eyes*.

Questions

a Is there any difference between using one eye and two? If so, try to explain this difference. What does this tell you about why we have two eyes?

b Is it easier or more difficult to use one hand or a partner to get the pencils opposite? If so, try to explain any difference.

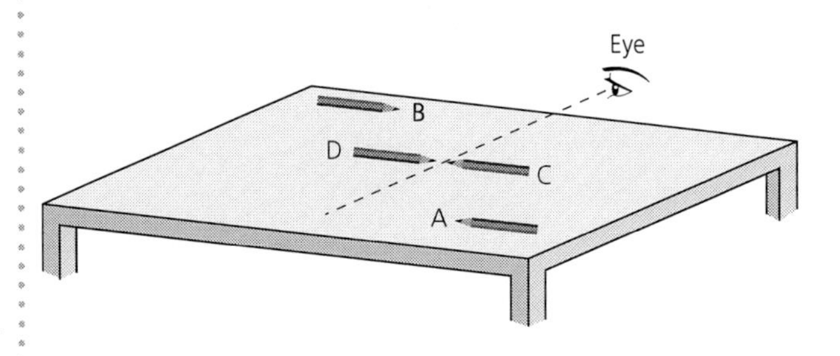

Three-dimensional vision

3-D vision allows you to see rounded, solid objects rather then a flat picture of your surroundings, like a photograph. This experiment helps you understand how your eyes and brain produce a 3-D vision.

1 Hold a large coin with its edge towards you, about 30 cm in front of your eyes.

2 Look at the coin first with you left eye closed and then with your right eye closed.

Questions

a What is the difference between the two views of the coin?

b Why does your brain need these two views to produce a 3-D vision?

c Try to think of reasons why, despite these results, your vision does not become completely flat and two-dimensional when you close one eye. (Do we need two eyes after all?)

Your teacher will be looking for:

- accurate observation
- good presentation of results and sensible conclusions

 Name:

Investigation of phototropism in cereal seedlings

You need:

- 4 pots containing germinated seedlings of wheat, or a similar cereal. The seedlings will probably require about 5 – 7 days before being used in the activity.
- Scalpel/razor blade (CARE)
- Forceps
- Aluminium foil
- Box made with lightproof material, but with a 2cm slit across its width at one end. Ideally the box should be lined with lightproof material, or made from black card.

Method

1 The pots of seedlings are labelled A, B, C and D, and treated as shown in the diagram.

2 Place the pots inside the box, with the labelled side of the pot facing the slit in the wall of the box. Make sure that the pots do not block out the light from one another.

3 Leave for one day near to a light source. This can be a natural one (e.g. a sunlit window), or an artificial one (e.g. a lamp).

4 After one day, remove the pots of seedlings from the box. Examine the seedlings and record your observations in a table like the one below.

Treatment of seedlings	Appearance after one day
A: tip removed	
B: tip covered with foil cap	
C: foil around coleoptiles (shoot), with tip exposed	
D: untreated	

Questions

1 How could you have made this investigation more reliable?

2 Suggest any steps which should be taken to make sure that your conclusions are valid.

3 Which part of the seedling coleoptiles *detects* light? Explain your answer.

4 Which part of the seedling coleoptiles *responds to* light? Explain your answer.

5 What is the name of the response you have observed? How does it help the plant to survive?

6 How would you expect a root to respond to light? Explain your answer. Explain how you could investigate this.

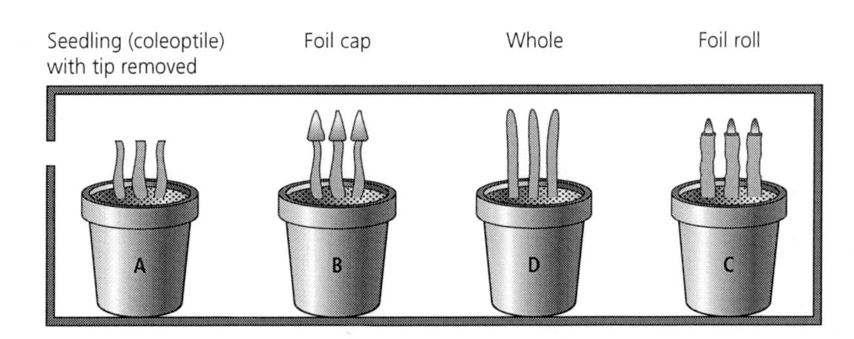

Name:

The structure of flowers

You need:

- mounted needle or long pin
- forceps
- scalpel
- hand lens
- white tile or paper
- examples of flowers (ideally one 'insect pollinated' and one 'wind pollinated')

Method

1 Examine each flower. Count the number of sepals and petals, if any are present. Remove these structures.

2 Examine the stamens (anther and filament). Remove them and record the number.

3 Cut horizontally across the female part of the flower. Count and record the number of carpels.

4 Collect one stamen from each of the flowers. Make a drawing of each of them in this space. Add a scale to your drawings.

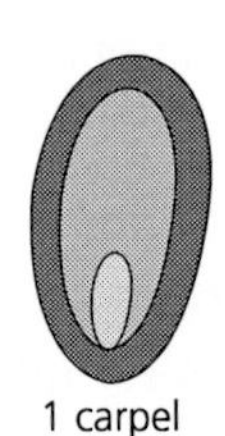

1 carpel

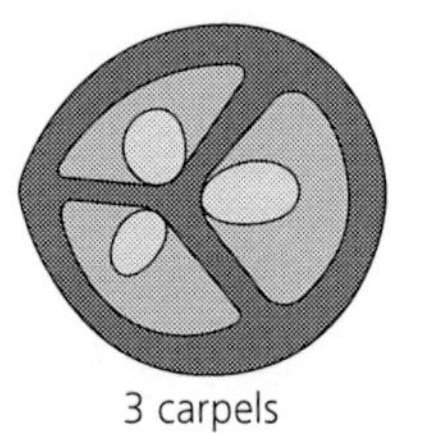

3 carpels

	No. of sepals	No. of petals	No. of stamens	No. of carpels
Insect-pollinated				
Wind-pollinated				

Questions

a What is the most obvious feature of the insect-pollinated flower that would be attractive to insects?

b Where, on the insect-pollinated flower, would you find the nectary? What is the purpose of this structure?

c How is the wind-pollinated flower adapted to its function?

Extension

If you have access to a source of ultra-violet light, shine this onto a petal from the insect-pollinated flower. What do you see? What is the significance of this?

 Name:

Conditions for germination of seeds

You need:

- five glass containers – test tubes, boiling tubes or jars
- cotton wool
- metal foil or black polythene
- cold boiled water
- marker pen
- liquid paraffin or light oil
- supply of small seeds, mung beans for example

Method

1. Set up the glass containers as shown, with ten seeds in each.
2. Add 2 or 3 drops of liquid paraffin/light oil to tube E – it will form a layer on the surface of the water.
3. Place containers A, B, D and E in a warm place, in the light, and place container C in a refrigerator or cold place. Leave for 48 hours.
4. Observe the seeds. Count and record the number that have split testas (seed coats). Once the testa has split the seed has started to germinate.
5. Calculate the percentage of the ten seeds that have begun to germinate in each tube, and complete the table below.

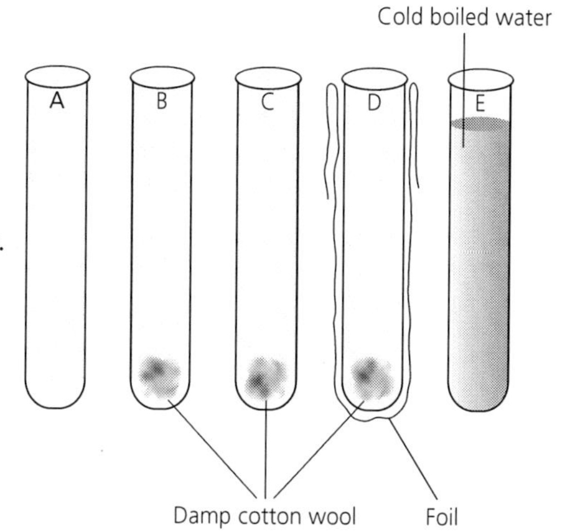

Questions

a. You could collect the results from other members of your class. How does adding together several sets of results improve the reliability of the data?

Tube	Factors present	Factor absent	Number of seeds germinating	Percentage germination
A	air, light, warmth	water		
B				
C				
D				
E				

b. You could modify the experiment by using single seeds e.g. broad bean. Would using single seeds affect the reliability of your results? Explain your answer.

c. Explain why each of the factors that you have identified is essential for seed germination.

Extension

Some seeds will not germinate unless they have passed through the gut of an animal. Use your knowledge of the process of digestion to explain why.

How could you investigate this in the laboratory? Be sure to identify independent, manipulated and controlled variables in your answer.

PRACTICAL **Name:** ....

Germination and growth

You need:

- cress and pea seeds
- paper towels
- 500 cm^3 beakers
- balances
- cotton wool
- test tubes and racks
- paraffin oil
- refrigerator

1 Warm light air
2 Warm water air
3 Cold water air
4 Warm water light air
5 Warm water light

Germination

1. Label five test tubes 1 to 5.
2. Put cress seeds in tube 1 and place it in a warm, well-lit place. These seeds have warmth, air and light but no water.
3. Put cress seeds on damp cotton wool in tube 2 and place it in a warm dark place. These seeds have warmth, air and water but no light.
4. Put cress seeds on damp cotton wool in tube 3 and place it in a refrigerator. These seeds have water and air but no warmth or light.
5. Put cress seeds on damp cotton wool in tube 4 and place it in a warm, well-lit place. These seeds have warmth, water, air and light.
6. Put cress seeds in tube 5 and cover them with boiled and cooled water (boiling drives oxygen out of the water), then pour a little paraffin oil onto the water to keep oxygen out. Place the tube in a warm, well-lit place. These seeds have warmth, water and light but no oxygen.
7. Examine the tubes after about three days. In which tubes have seeds germinated? What conditions are necessary for seeds to germinate?

Growth curves

If you weigh organisms as they grow and plot your results on a graph the result is a curved shape called a **growth curve**. Plot two different growth curves as follows.

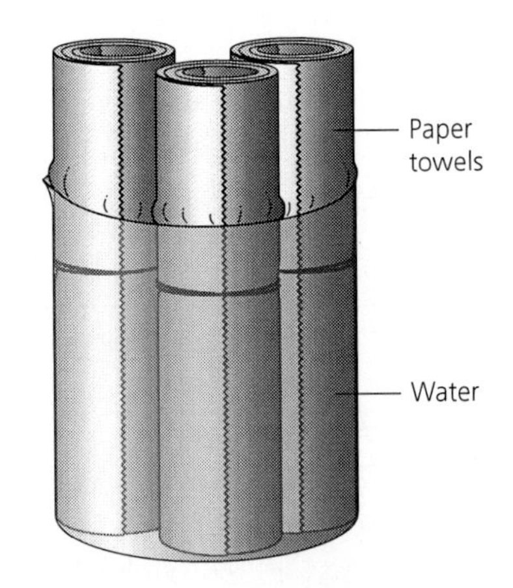

1. Soak 40 peas in water for 12 hours, then wrap them in wet paper toweling. Put the wrapped peas in beakers of water in a warm, dark place (see diagram opposite).
2. Every three days remove five seedlings and find their average weight. Then heat them at 100° C until they are completely dry and find their average weight again. Plot graphs for wet and dry weight changes.
3. Explain why the wet and dry growth curves are different. What conclusions can you draw about the early growth of plants from these results?

Your teacher will be looking for:

- careful use of the apparatus given
- accurate observations and measurements
- good presentation of results including tables and graphs
- sensible conclusions which fit your results

 Name:

A model for genetics

You need:

- some coins of the same denomination
- circular sticky labels which fit a coin

H and **h** are alleles of the gene for hair colour.

H is dominant and if it is present the hair is dark. If both alleles are **h** the hair is fair.

		ova	
		H	h
sperms	H	HH	Hh
	h	hH	hh

Look at the diagram above showing the zygotes which could be produced by a father with the genotype **Hh** and a mother with the genotype **Hh**.

The diagram shows that zygotes with a **dominant allele** (**HH**, **Hh** or **hH**) are three times more likely to be produced than zygotes with two **recessive alleles** (**hh**). In other words, dominant and recessive phenotypes occur in the ratio of 3:1.

This happens because:
- half the sperms carry the **H** allele and half carry the **h** allele
- half the ova carry the **H** allele and half carry the **h** allele
- there is an equal chance that, during fertilisation, any sperm can fertilise any ovum (i.e. fertilisation is a **random process**).

This can be checked experimentally using coins to represent sperms and ova, and coin tossing to represent the random process of fertilisation.

Method

1 Work in groups of two. Each group should obtain two coins. Use circular sticky labels to mark one side of one coin **sperm A** and the opposite side **sperm a**, then label one side of the other coin **ovum A** and the opposite side **ovum a**. Copy the chart below.

2 Working in pairs, spin the 'sperm coin' and the 'ovum coin' at the same time. Look at how they fall and enter the result in the appropriate part of the tally column. Repeat at least 50 times, then work out the totals.

3 Does the ratio of dominant to recessive phenotypes come to about 3:1?
Why must the coins be tossed at least 50 times to get reliable results?

Sperm coin

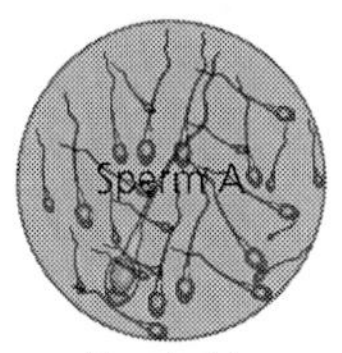

Head side

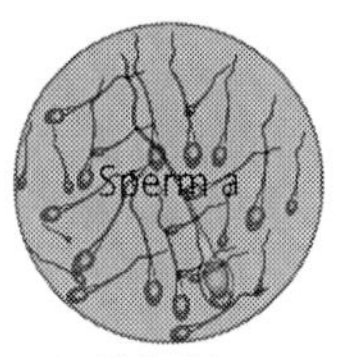

Tail side

Ovum coin

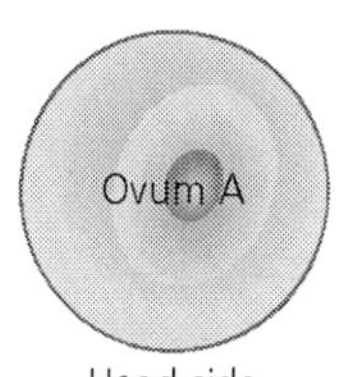

Head side

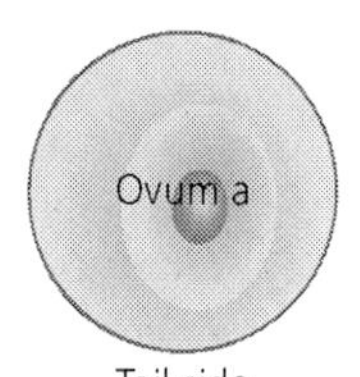

Tail side

sperm	ovum	tally	
A	A		total = (A A, a A, A a)
a	A		
A	a		
a	a		total =

Further work

B and **b** are the alleles for eye colour. **B** is dominant and if present the eyes are brown. If both alleles are **b** the eyes are blue. A blue eyed man has the genotype **bb**. He is married to a brown eyed woman with the genotype **Bb**.

- What are the chances that their first child will have blue eyes?
- What are the chances that their second child will have blue eyes?
- If they have four children, how many will have brown eyes? (Be careful – can you be **sure**?)

 Name: ....

Looking at variation

You need:

- tape measure
- scales (in kilograms)
- ruler with millimetre scale
- graph paper

1 Working in pairs, record the information listed on the next page for both students, referring to the diagrams if necessary.

2 Record your results.

Data for the whole class should be recorded in the form of a table on the chalkboard (see opposite).

Data should be used to draw histograms.

A blank histogram for finger length is shown opposite.

Draw others for the remaining data.

3 For each of the characteristics which you have measured/observed, decide whether the particular feature is an example of continuous variation or discontinuous variation. Briefly explain your reasons for deciding whether the characteristics were continuous or discontinuous.

4 Refer specifically to the data on hand span and height. Plot a graph which would enable you to determine whether these two characteristics are related. Comment on your results.

name	ear lobes yes/no	tongue rolling yes/no	finger length (mm)	mass (kg)

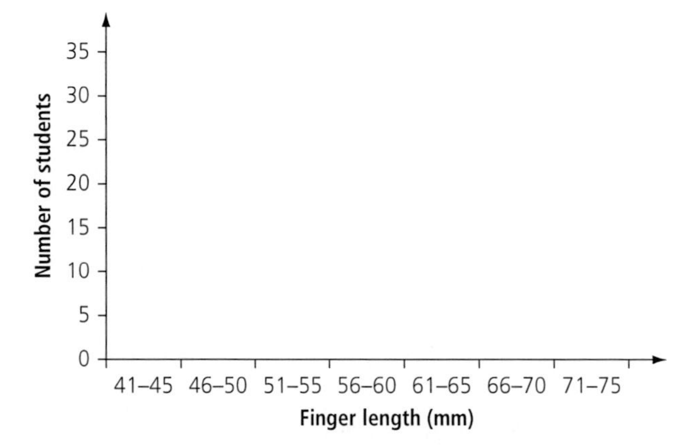

Your teacher will be looking for:

- careful measurement and recording of results
- good presentation of results in tables and histograms

Further work

- Think of other variations you could investigate for humans. For each one, predict whether it is continuous or discontinuous.
- Think of variations you could study in plants. Investigate one of these by collecting data and presenting your results as a histogram.

 Name: ....

Human variation

You will be asked to make a number of simple observations on yourself, and then to pool your observations with the other members of the form. Once the observations have been made, you will be asked a number of questions about them.

1 Ears: are your ears lobed or unlobed (diagram A)?

Lobes?..................................

2 Thumbs: clasp your hands together, with the fingers interlocking.

Which thumb is on top?..................................

3 Tongue-rolling: can you roll your tongue with a 'U' down the middle? (diagram B)

Are you a roller or a non-roller?........................

4 Index-finger: measure the length of your index finger, from the tip to the junction with the palm of the hand (diagram C).

Length of index finger?.................................

5 Pulse rate: record your pulse rate, whilst sitting after five minutes' rest.

Pulse rate at rest?....................................

6 Thumb shape: hold your hands in front of you, with the thumbs spread out. Are your thumbs fairly straight, or do they curve like those of a hitch-hiker?

Thumb shape?..

7 Big-head: use a tape measure to determine your cranial circumference at the level of your forehead.

Cranial circumference?..................................

8 Hand span: measure the maximum distance between the tip of your right thumb and the tip of your right little (fifth) finger.

Hand span?..

9 Height: measure your height, in stockinged feet

Height?...

Now collect the same data from all the other members of the form. Present these data in a suitable form.

A Ears

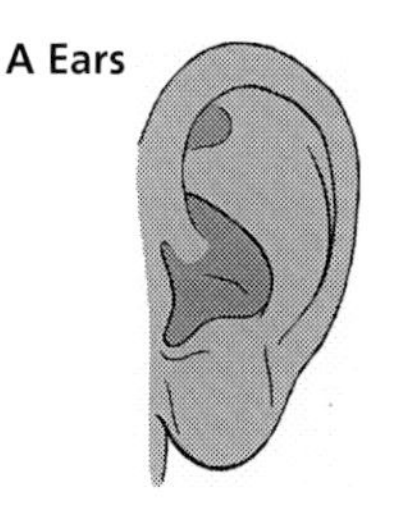

ear lobe present

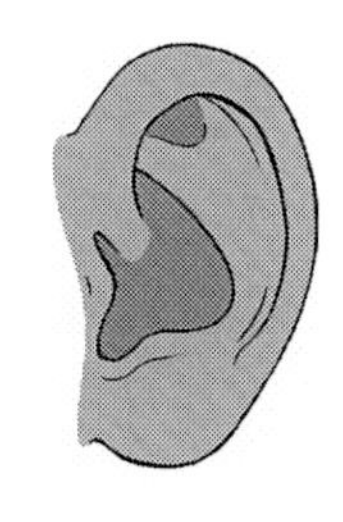

ear lobe absent

B Tongue
rollers or non–rollers

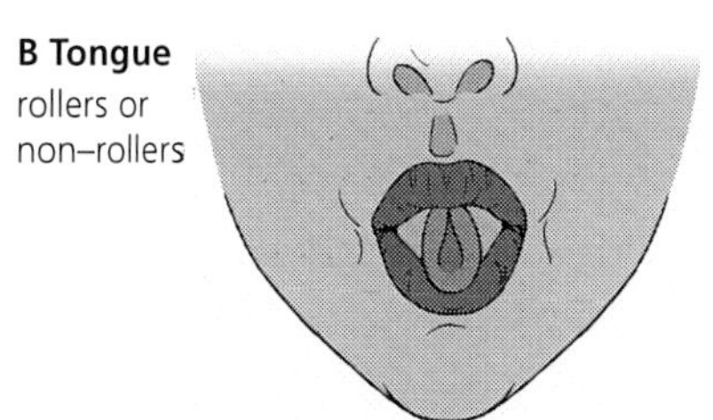

C Finger length (mm)

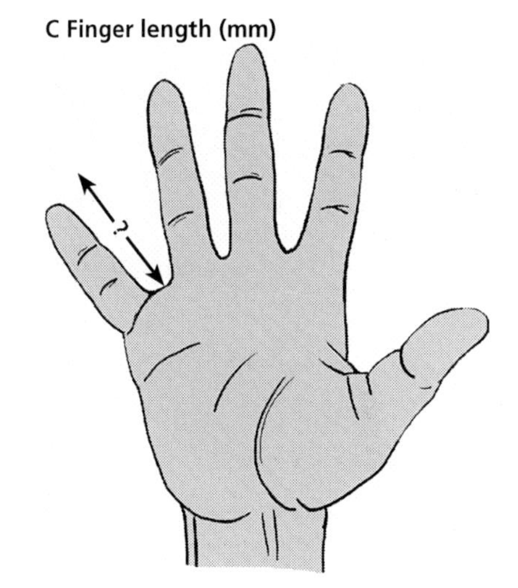

PRACTICAL **Name:**

Decomposition of cellulose by bacteria

You need:

- tubes/containers of sterile salt solution and filter paper
- 10 cm^3 measuring cylinders
- 100 cm^3 measuring cylinder
- spatula
- soil sample
- marking pen
- ruler
- distilled water
- metal foil

Paper is made from crushed and pulped wood – the plant cells have walls made up of cellulose.

Method

1. Label the tubes A, B and C.
2. Put a 1 cm depth of soil in the 10 cm^3 cylinder. Add water to make the volume up to 10 cm^3. Shake the mixture.
3. Add 10 drops of this soil/water mixture to tube A. Try not to drip the mixture onto the filter paper in the tube. Seal with foil.
4. Put a 1 cm depth of soil in the second 10 cm^3 measuring cylinder. Transfer the soil to the 100 cm^3 cylinder. Add water to make the volume up to 100 cm^3. Shake the mixture.
5. Using a clean dropper, add 10 drops of this soil/water mixture to tube B. Seal with foil.
6. Using a clean dropper, add 10 drops of distilled water to tube C. Seal with foil.
7. Keep the test-tubes in a warm place for 14 days.
8. After 14 days, record the appearance of the filter paper near the air/water junction.
9. Shake the tubes. Record what happens to the filter paper.

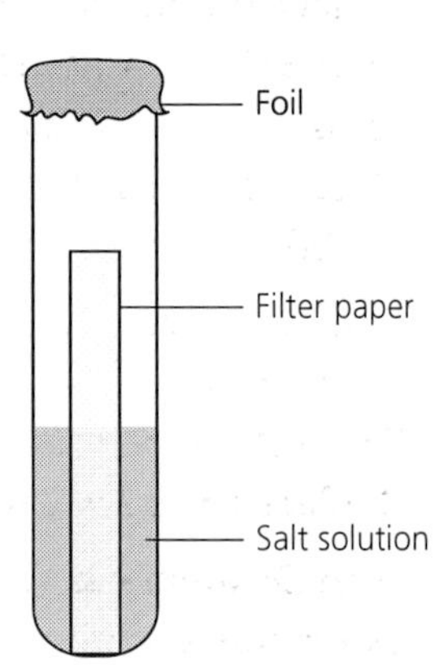

Tube	Soil added	Appearance of filter paper	What happened on shaking
A	$\frac{1}{10}$		
B	$\frac{1}{100}$		
C	none		

Questions

a. If the filter paper falls apart on shaking, what does this show?
b. What was the purpose of tube C?
c. What is the importance of cellulose decomposition in the recycling of natural materials?

Extension

A yellow or orange stain on the filter paper indicates that cellulose has been digested. Which enzymes would be needed to decompose cellulose?

How could you test that the cellulose had been broken down to its subunits?

Humans build many wooden structures. How can we prevent this decomposition from damaging our buildings?

Animals cannot make cellulose-digesting enzymes. Find out how cattle and termites are able to digest cellulose.

 Name: ....

The effects of soap and pH on bacteria

You need:

- Petri dishes of sterile nutrient agar and tubes of nutrient broth
- sterile bulb pipettes
- tubes of nutrient broth
- incubator
- bacteria culture
- warm water, soap, and paper towels

How clean are your hands?

1 Label four Petri dishes of sterile nutrient agar A, B, C, and D.

2 **Dish A:** Take off the lid for 10 seconds, then replace it and seal the. rim with sticky tape.

Dish B: Take off the lid and press the fingers of an *unwashed* hand onto the agar (it must not be broken up by too much pressure). Replace the lid within 10 seconds and seal it.

Dish C: Wash your hands in warm water only (no soap). Dry them with paper towels. Take off the lid and touch the agar as before. Replace the lid within 10 seconds and seal it.

Dish D: Wash your hands *thoroughly* using warm water and soap. Dry them with paper towels. Take off the lid, touch the agar as before, replace the lid within 10 seconds and seal it.

3 Incubate the Petri dishes upside down at 30° C for a week. *Without opening them:*

a count the number of bacteria colonies in each dish

b count the number of different colonies in each dish

c design a results table

What do your results tell you about the cleanliness of washed/unwashed hands and the effectiveness of soap as a cleaning agent?

Your teacher will be looking for: **and especially for:**

- careful and safe use of the apparatus given ☐
- accurate observation and recording of results ☐
- good presentation of results in tables ☐
- sensible conclusions ☐
- .... ☐

Investigate the effects of pH on bacterial growth

4 Add 8 cm^3 of sterile nutrient broth to each of five sterile boiling tubes marked A, B, C and D.

5 Add 1 cm^3 of 0.1M hydrochloric acid to tube A.

Add 1 cm^3 of 0.0001M hydrochloric acid to tube B.

Add 1 cm^3 of distilled water to tube C.

Add 1 cm^3 of 0.0001M sodium hydroxide to tube D.

Add 1 cm^3 of 0.1M sodium hydroxide to tube E.

6 Inoculate each tube with 1 cm^3 of bacterial culture, plug them with cotton wool and seal with sticky tape.

Incubate the tubes at 25° C to 30° C for 48 hours.

Compare the cloudiness (**turbidity**) of each tube. What do your results tell you about the effect of Ph on bacterial growth?

Further work

a A hospital wants to test a new bactericidal soap for cleaning the hands of surgeons. Design an experiment to compare the new soap with the one they already use.

b Onions and other vegetables can be preserved in weak acids such as vinegar. Design an experiment to find the strength of vinegar required to preserve vegetables at room temperature for at least one month.

HAZARD WARNING

Wash hands thoroughly after touching agar. During incubation, plates SHOULD NOT be completely sealed. After incubation, plates must be completely sealed with tape before observing results.

PRACTICAL **Name:** ....

Ecology – studying populations

You need:

- 25 cm quadrats
- graph paper
- enamel paint
- paint solvent
- fine paint brush
- notebooks
- identification books (plant and animal)
- white pie dishes
- specimen tubes

Random sampling of plant life

How would you answer the question, 'Which types of plant are most common in this habitat?' You could count all the different types of plant but this is not necessary except for very small habitats. An easier method is called **random sampling**. To do this you use a square or rectangular frame called a **quadrat** to study several small areas (**samples**) of the habitat chosen at random.

You place a quadrat at random throughout a habitat by throwing one over your shoulder. Take care! Do not deliberately throw it to land on vegetation which looks interesting. What you do next depends on the information you require. This could be the *density, frequency* or *percentage cover* of various types of plant.

1 **Density** This is the number of plants (or animals) in a unit area of habitat (e.g. the number per 25 centimetre square). To discover the density of a plant species in a habitat, you count the number of this species present inside the area of a quadrat each time it lands. Continue until the quadrat has been cast throughout the whole habitat, then calculate the average number of times the species was found.

2 **Frequency** This is the number of times that a particular species is found when a quadrat is thrown a certain number of times. To calculate frequency you count the number of different species within the quadrat each time it lands and note their names. If, for example, you throw it a hundred times, you note the number of times each species was found and express each result out of a hundred. This will tell you the most common (most frequent) species in the habitat, then the next most common, down to the rarest.

A simple quadrat

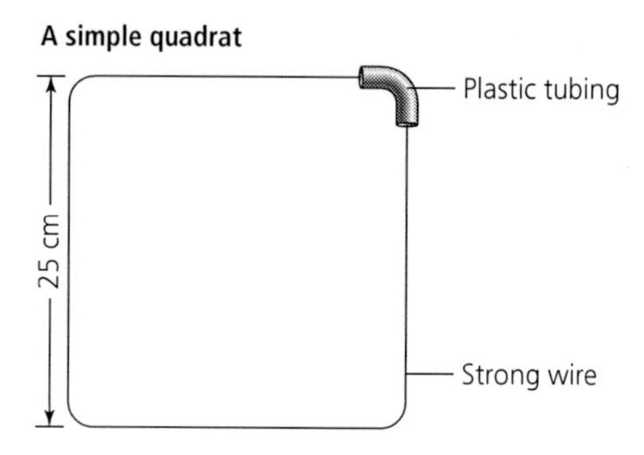

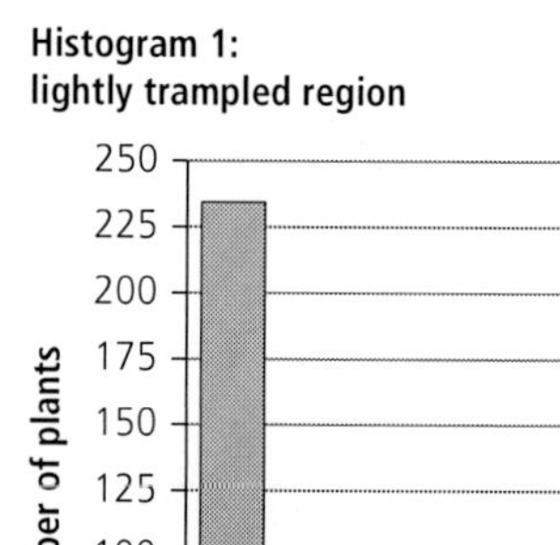

Histogram 1: lightly trampled region

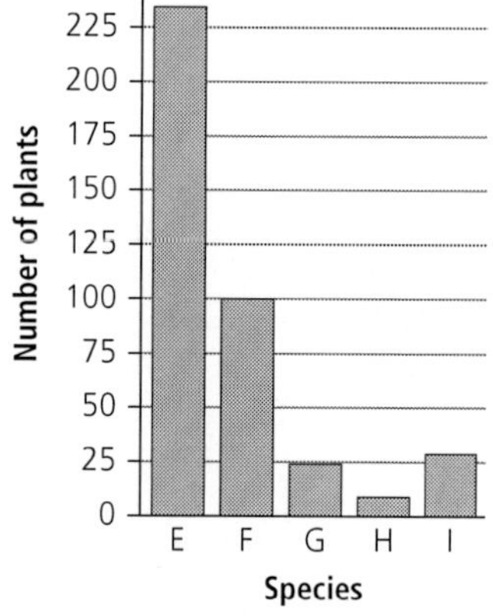

Histogram 2: heavily trampled region

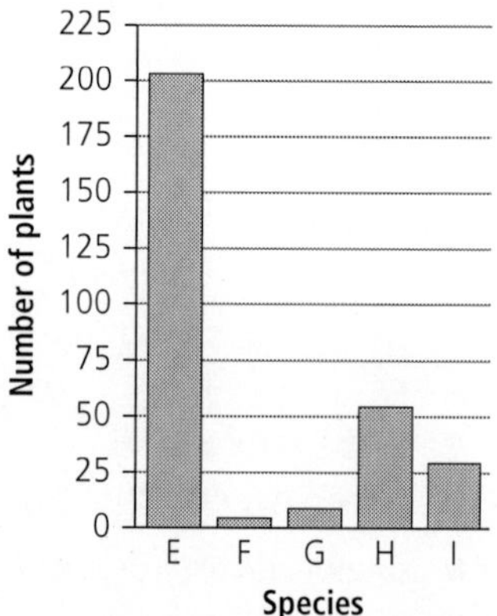

PRACTICAL **Name:** ....

Handling experimental observations and data: estimating the size of a population

You need:

- plain paper
- pen or pencil
- bag or small box

Method:

You are provided with a bag with some pieces of paper in it. The pieces of paper represent animals in a population, and the bag is the environment in which they live. The investigation looks at a 'capture-recapture' technique for estimating their population size.

1. Remove between 15 and 20 'animals' from the habitat (the exact number does not matter) and record this number in the table.
2. Mark all the pieces of paper with a small number 1 and put them back in the bag. Shake for 1 minute to mix up the 'animals'.
3. Remove 15–20 'animals' from the bag and write down this number in the table.
4. Count how many of this second sample have got a number 1 written on them (remember to look on both sides).
5. Estimate the size of the population using this formula:

$$\frac{\text{No. in first sample} \times \text{no. in second sample}}{\text{No. in second sample marked with a 1}}$$

6. Repeat steps 1–5 a further 4 times but mark the captured 'animals' in the second step with a 2 the second time, a 3 the third time, a 4 the fourth time and a 5 the fifth time. Ignore any other numbers from the earlier samples.
7. Display all the readings in the form of a table. Work out a mean value for the estimate of the population using your five sets of results. Record the mean value.
8. Tip out all of the 'animals' and count the actual population size. Record this value.
9. Present your results in a bar chart that shows all five of the estimates of the population and the actual value clearly.
10. Calculate the percentage error in your estimates compared with the actual value.

 Name:

Browning of apples and pH

Introduction

Freshly cut apples gradually turn brown when they are left in air. Cooks try to stop this by dipping the slices of apple in lemon juice.

Design and carry out an investigation to find out whether pH affects the rate of the 'browning' reaction.

Start by thinking about the following:

- How will you prepare solutions with a range of pH values?
- What apparatus will you need and how will you use it?
- How will you decide when the apple pieces have turned brown?
- Will you need to set up any controls?
- Write down any *hypotheses* you are going to test.
- Are there any safety hazards?
- What safety precautions **must** you take?

Plan your investigation

Let your teacher check your plans

Carry out your investigation

Write up:

- What you did (including diagrams).
- What you found (your results including any tables).
- What your conclusions are.
- Whether your hypothesis was supported (proved) or not.
- Any scientific explanation you can offer for your conclusions.
- How your investigation could be improved.

Your teacher will be looking for:

- use of a sensible method
- use of solutions with a sensible range of pH values
- careful observations and measurements
- good presentation of results
- sensible conclusions
- sensible suggestions about improving the experiment

Name: ....

Investigating fermentation

Introduction

Fermentation uses yeast to convert sugars into alcohol. It is one of the oldest uses of biotechnology. As early as 6000 BC the Babylonians were brewing beer and by 4000 BC the Egyptians were using yeast to make bread rise.

Design and carry out an investigation into the factors affecting the rate at which fermentation takes place.

Fermentation – useful biotechnology

Start by thinking about the following:

- What do you know about fermentation?
- What apparatus will you need and how will you monitor the rate of fermentation?
- What factors will you investigate? How will you keep other factors constant?
- How many experiments will you need to carry out?
- Write down any *hypotheses* you are going to test.
- Are there any *safety hazards*?
- What safety precautions **must** you take?

Plan your investigation

Let your teacher check your plans

Carry out your investigation

Write up:

- What you did (including diagrams).
- What you found (your results including any tables and graphs).
- What your conclusions are.
- Whether your hypotheses were supported (proved) or not.
- Any scientific explanation you can offer for your conclusions.
- How your investigation could be improved.

Your teacher will be looking for:

- use of sensible method
- good choice of factors to investigate
- careful observations and measurements
- good presentation of results
- sensible conclusions which fit your results
- sensible suggestions about improving the experiment

INVESTIGATION **Name:** ....

Investigating artificial meat

Introduction

Many people choose not to eat meat. Some do not think that it is healthy and others do not like the way animals bred for meat are treated.

However, it is now possible to buy 'artificial meat'. Some is made from the soya bean plant *(Glycine soja)*. Another type is made from a fungus called *Fusarium* which can grow on potatoes, starch, or wheat. The artificial meat is processed to make it look, feel, and taste like real meat.

Design and carry out an investigation to compare real meat with artificial meat.

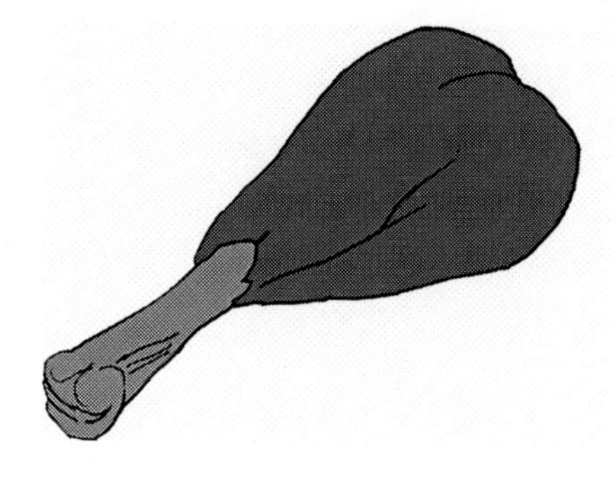

Real or artificial

Start by thinking about the following:

- What can you find out about artificial meat?
- What do you know about food test? What food groups will you test for in this investigation?
- How are you going to make sure that your tests are 'fair'?
- How will you display your results?
- Are there any safety hazards?
- What safety precautions **must** you take?

Plan your investigation

Let your teacher check your plans

Carry out your investigation

Write up:

- What you did.
- What you found (your results including any tables).
- What your conclusions are.
- How your investigation could be extended to include factors such as taste.

Your teacher will be looking for:

- a sensible experimental design
- safe use of apparatus and chemicals
- careful and accurate observations
- good presentation of results
- sensible conclusions
- sensible suggestions about extending the experiment

Name:

How safe to refrigerate?

Introduction

Many foods have to be kept in a refrigerator to stop them going off too quickly. Even in a refrigerator, food can only be kept for a certain length of time before it is unsafe to eat. Sometimes this information is given on the food label.

Design and carry out a scientific experiment to find out how long it is safe to keep food in a domestic fridge.

You should investigate food kept in the fridge all the time and food taken out for half an hour each day and then put back again (as could happen, for example, with cooked meat).

KEEP REFRIGERATED AFTER OPENING EAT WITHIN FOUR DAYS

Start by thinking about the following:

- Why does food go bad? Why does this make it unsafe to eat?
- What have you studied about 'bacteria'? Can you use this knowledge?
- What types of foods do you think will be best to use?
- What do you *predict* will happen? Why?
- How will you check for bacteria? (Visible checking is not good enough.)
- What safety precautions must you take?
- Are there any safety hazards?
- What safety precautions **must** you take?

Plan your investigation

Let your teacher check your plans

Carry out your investigation

Write up:

- What you did (include diagrams).
- What you found (your results including any tables).
- What your conclusions are and how they relate to your predictions.
- How your experiment could be improved.

Your teacher will be looking for:

- the design of a fair test
- the use of proper safety precautions
- careful observations and measurements
- good presentation of results
- a sensible conclusion
- sensible suggestions about improving the experiment

Name: ....

Light and photosynthesis

Introduction

When plants photosynthesise, they produce oxygen. When water plants give off oxygen, small bubbles of the gas can be seen. A simple way of measuring the rate of photosynthesis is to count the bubbles released in a certain period of time.

Design and carry out a scientific experiment to find out whether the intensity of light affects the rate of photosynthesis in the pond weed *Elodea*.

Start by thinking about the following:

- What do you know about photosynthesis?
- What apparatus will you need to study photosynthesis in *Elodea*?
- How will you vary light intensity? Will you have to set up any controls?
- What do you *predict* will happen? Why?
- Are there any safety hazards?
- What safety precautions **must** you take?

Plan your investigation

Let your teacher check your plans

Carry out your investigation

Write up:

- What you did (include diagrams).
- What you found (your results including any tables and/or graphs).
- What your conclusions are and how they relate to your predictions.
- How your experiment could be improved.

Your teacher will be looking for:

- the use of a sensible scientific method
- careful observations and measurements
- good presentation of results
- sensible conclusion
- sensible suggestions about improving the experiment

 Name: ....

Investigating plants and soil acidity

Introduction

Some plants grow better in soils which are slightly acid (pH < 7) whilst others prefer slightly alkaline soils.

First do some research to find the names of common garden plants which are 'acid-haters' and those which are 'acid-lovers'.

Then find examples of those plants growing in gardens or even pots. Investigate the acidity of the soil in which they are growing in order to test the information you obtained in your research.

Cabbages and other brassicas prefer a slightly alkaline soil. Prepare the vegetable bed by adding a light dusting of lime before planting.

Start by thinking about the following:

- What do you know about testing soil pH? What will you need?
- Write down the *hypothesis* you are going to test for each plant.
- Who may be able to help you find suitable plants?
- How are you going to decide whether a plant is growing well or not?
- How many plants will you need to investigate to test your hypotheses?
- Are there any safety hazards?
- What safety precautions **must** you take?

Plan your investigation

Let your teacher check your plans

Carry out your investigation

Write up:

- What you did.
- What you found (your results including any tables).
- What your conclusions are.
- Whether all your hypotheses were supported (proved) or not.
- How your investigation could be improved and/or tested in the laboratory.

Your teacher will be looking for:

- a sensible choice of plants
- careful observations and measurements
- good presentation of results
- sensible conclusions
- sensible suggestions about improving the experiment

 Name: ....

Improving garden soil

Introduction

Some plants grow better in soils which hold the water so that the roots are always surrounded by moist earth. Others prefer quick draining soils.

Gardeners try to make ideal soils for a wide range of plants by mixing ordinary garden soil with sharp sand and/or peat.

Design and carry out an investigation into the effect of mixing garden soil with:

a sharp sand or grit

b peat or coir

Cacti like free-draining soils

Start by thinking about the following:

- What do you know about testing soil for 'water retention' and 'permeability'?
- What apparatus will you need and how will you use it?
- How many mixtures of soil will you test? How will you control their compositon?
- Write down any *hypotheses* you are going to test.
- Are there any safety hazards?
- What safety precautions **must** you take?

Ferns like moist soils

Plan your investigation

Let your teacher check your plans

Carry out your investigation

Write up:

- What you did (including diagrams).
- What you found (your results including any tables).
- What your conclusions are.
- Whether your hypotheses were supported (proved) or not.
- Any scientific explanation you can offer for your conclusions.
- How your investigation could be improved.

Your teacher will be looking for:

- use of a sensible method
- use of a sensible range of soils careful prepared
- careful observations and measurements
- good presentation of results
- sensible conclusions
- sensible suggestions about improving the experiment

INVESTIGATION **Name:**

The effect of mineral deficiency on the growth of seedlings

You need:

- 6 boiling tubes and rack
- Black polythene, black paper or aluminium foil
- Elastic bands
- Cotton wool
- Distilled water
- Complete culture solution/ nitrogen-deficient culture solution/ calcium-deficient nutrient solution/ phosphate-deficient solution/ magnesium-deficient solution.
- (A mixture to prepare Sach's water culture solutions is available from Philip Harris. It can be cheaper (and is certainly much easier) to buy the ready-prepared nutrient solutions if not all the solid compounds are available in the school. Your own solutions can be prepared, if desired, using recipes from CLEAPSS Recipe card 73).
- 6 seedlings of maize or a similar cereal. The seeds should be moistened to germinate about a week before the activity is planned.
- Ruler/scalpel or razor blade (CARE)/ waterproof ink/fine paintbrush

Method

1 Wrap polythene, paper or aluminium foil around each of the boiling tubes.

2 Label the tubes A – F: fill tubes as follows.
A – distilled water, B – complete culture solution,
C – calcium-deficient solution, D – nitrogen-deficient solution,
E – magnesium-deficient solution, F – phosphate-deficient solution.

3 Carefully trim the endosperm food store away from each of the cereal seeds, then insert the maize seedlings into a strip of cotton wool as shown in the diagram. Measure the length of the main root of the seedling, and mark the root with a ring of waterproof ink (this will allow you to check which was the longest root at the start of the activity). Place the wrapped seedling in the top of tube A so that its roots extend into the distilled water.

4 Repeat step 3 for each of the tubes B – F.

5 Leave the seedlings in a light, airy position for 3 weeks. Check the level of the solutions each week, and top up with distilled water if necessary.

6 After 3 weeks, examine each of the seedlings and record results in a table like the one below:

Tube	Solution	Length of longest root at day 1(mm)	Length of longest root after 21 days (mm)	Change in root length (mm)	Number of roots	Colour and condition of leaves
A						
B						
C						
D						
E						
F						

Questions

1 Why were the boiling tubes wrapped in polythene, paper or aluminium foil?

2 Why was the food store trimmed away from the seedling at the start of the activity?

3 Why is nitrogen required for plant growth?

4 What is the main effect of a deficiency of magnesium?

5 Describe how you could adapt this method to investigate the effects of mineral deficiency on a floating plant, such as duckweed (*Lemna* species).

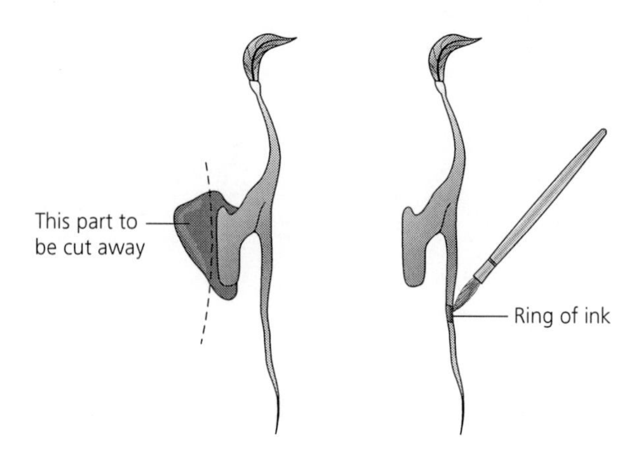

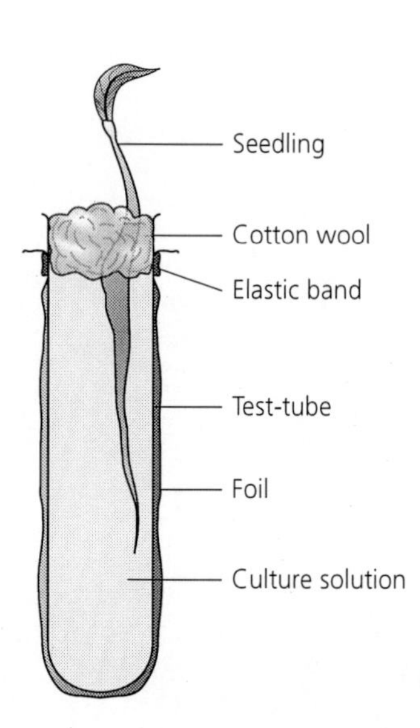

INVESTIGATION **Name:** ....

Transpiration through leaves

Introduction

When plants transpire, water is lost through their leaves.

Design and carry out an investigation to decide whether transpiration takes place through the upper surface of the leaf, the lower surface, or both!

Start by thinking about the following:

- What do you know about transpiration?
- How will you detect water loss?
- What apparatus will you need and how will you use it?
- How many leaves will you test? Will they need to be on the plant or not?
- Write down any *hypotheses* you are going to test.
- Are there any safety hazards?
- What safety precautions **must** you take?

Plan your investigation

Let your teacher check your plans

Carry out your investigation

Write up:

- What you did (including diagrams).
- What you found (your results including any tables).
- What your conclusions are.
- Whether your hypotheses were supported (proved) or not.
- Any scientific explanation you can offer for your conclusions.
- How your investigation could be improved or extended.

Your teacher will be looking for:

- use of a sensible method
- use of a sensible number of leaves carefully prepared
- careful observations and measurements
- good presentation of results
- sensible conclusions which match your results
- sensible suggestions about improving the experiment

 Name:

Investigating respiration rates

Introduction

When animals respire aerobically, they use oxygen (usually from the air) and produce carbon dioxide gas. By monitoring how fast they use oxygen (and produce CO_2) we can measure their respiration rate.

Design and carry out an investigation into the effect of temperature on the respiration rate of maggots or other invertebrates.

$$\text{oxygen} + \text{glucose} \downarrow \text{carbon dioxide} + \text{water} + \text{energy}$$

Start by thinking about the following:

- Have you seen any apparatus suitable for measuring respiration rates? If not, try to find a suitable method in a textbook (or this book!).
- What range of temperatures will you use? How will you control the temperature?
- Are there any other factors which you must keep constant?
- What do you *predict* will happen? Write down any *hypotheses* you are going to test.
- Are there any safety hazards?
- What safety precautions **must** you take?

Plan your investigation

Let your teacher check your plans

Carry out your investigation

Write up:

- What you did (including diagrams).
- What you found (your results including any tables).
- What your conclusions are.
- Whether your hypothesis was supported (proved) or not.
- Any scientific explanation you can offer for your conclusions.
- How your investigation could be improved.

Ethics in biology

Remember you must not injure or kill the maggots, or cause them any harm.

Your teacher will be looking for:

- use of a sensible method
- use of a sensible range of temperatures carefully controlled
- careful observations and measurements
- good presentation of results
- sensible conclusions
- sensible suggestions about improving the experiment

 Name:

Fitness and lung volume

Introduction

Athletes have to produce energy quickly. It helps if they can get oxygen to their muscles as it is needed. If not, an oxygen debt builds up, the muscles start to respire anaerobically and fatigue sets in.

Design and carry out an experiment to see if there is a relationship between lung volume and 'fitness'.

Start by thinking about the following:

- How can you measure lung volume?
- What do you mean by fitness? Is it about power or is it about recovery rate? Or can you think of a better definition?
- How many people will you need to test?
- Are there any other factors you should take into account?
- Write down any *hypotheses* you are going to test.
- Are there any safety hazards or hazards to health?
- What safety precautions **must** you take?

Plan your investigation

Let your teacher check your plans

Carry out your investigation

Write up:

- What you did (including diagrams).
- What you found (your results including any tables).
- What your conclusions are.
- Whether your hypotheses were supported (proved) or not.
- Any scientific explanation you can offer for your conclusion.
- How your investigation could be improved.

Your teacher will be looking for:

- use of a sensible method for measuring fitness
- use of a sensible method for measuring lung volume
- careful observations and measurements
- good presentation results
- sensible conclusions
- sensible suggestions about improving the experiment

Answers to student book questions

1.1 **1** About 2.2 billion years

2 Amino acids, fatty acids/lipids, sugars (nucleotides/nucleic acids)

3 REDGIRL, MRSGREN for example

4 To replace individuals that die/provide more individuals to colonise new environments/to combine genetic material from different individuals (provide variation)

1.2 **1** **a** Have well-developed carnassials (cutting) teeth

b Weasel and mink (same genus)

c The fox (the others are in the same family)

d They ingest their organic molecules in ready-made form

2 The wise man

1.3 **1** Several possibilities e.g. begin with spherical/not spherical, then 'not spherical' could be with tails or flagella/without tails/flagella

2

	Animal	Plant	Bacterium
Cell wall	–	+	+
Slime capsule	–	–	+
Cell membrane	+	+	+
Nucleus	+	+	–
Chloroplast	–	+	–
Mitochondrion	+	+	–
DNA	+	+	+
cytoplasm	+	+	+

3 Because they do not show the typical characteristics of living organisms, unless they are inside the cells of another living organism

4 10 240

1.4 **1** Dark means little light for photosynthesis, so fungi can out-compete the green plants as fungi obtain food without photosynthesis

2

Enzyme	Substrate	Product
Amylase	Starch	Maltose
Lipase	Fat/lipid	Fatty acids and glycerol
Protease	Protein	Peptides/amino acids

3 Could consume human food – e.g. mould on bread
Could cause disease – e.g. athlete's foot
Could destroy human's buildings – e.g. wooden buildings in damp areas
Can be used in biotechnology – e.g. in baking and brewing

1.5 **1** Algae, mosses, ferns, seed plants. Algae do not have a waterproof cuticle, and they have no vascular system to transport water from soil around their bodies.

2 Their gametes can be transferred without dependence on water

1.6 **1** they ingest ready-made organic molecules; a backbone; annelids; chaetae; molluscs; arthropods

2 **a** They have segmented bodies/segments carry jointed limbs/they have a hard exoskeleton

b Two pairs of wings/three pairs of legs/three major body sections/ metamorphosis in life cycle

c Insects: 6, 3, 4, compound; spiders: 8, 2, none, simple

3 Life cycle has four distinct stages: egg, larva, pupa/chrysalis, adult. Adults can larvae can lead different lives, on different food sources – reproductive structures can develop during stage as pupa.

2.3 **1** A structure which allows some molecules/particles to cross but prevents others from crossing. For example, the membrane lining the ileum allows glucose and amino acids to cross but prevents starch and large proteins from crossing.

2 A 'slope' between a high and a low concentration of a dissolved substance

3 Movement of gases (e.g. oxygen and carbon dioxide), food substances (e.g. glucose), wastes (e.g. urea), hormones (e.g. insulin)

4 Large surface area/thin membranes (short pathway for diffusion)/ability to set up and maintain diffusion gradients

5 cytoplasm; partially-permeable membrane; swell; lower; cell wall; glucose/amino acids; diffusion; concentration gradient; active transport; energy; against; carbon dioxide; diffusion; photosynthesis

6

	Diffusion	**Active Transport**
Is energy(ATP) used	no	yes
Direction of movement	With/down a concentration gradient	Can be against a concentration gradient
Are carrier molecules involved	No	Yes
Example	Oxygen movement from blood to air sacs	Uptake of mineral ions by root hair cells

Phagocytosis would be used to take up/move large particles across a membrane

2.4 **1** **proteins;** reactions/processes; catalysts; specific; denatured

2 Draw diagram in reverse (i.e. one molecule on active site becoming two products): any digestive enzyme would be a suitable example

2.6 **1** **a** 0.25/0.50/2.0/0.5/0.25/0.08

b Temperature on x axis, rate on y axis, maximum rate at 35°C

c Up to 35°C rate increases as molecules gain more kinetic energy and so collide more often and with greater energy, but above this temperature the enzyme molecules begin to denature. Active site no longer 'recognises' substrate molecules and so rate decreases. 35°C is the optimum temperature for this enzyme.

2.7 **1** Carbohydrates (source of energy)/lipids (energy store)/proteins (enzymes)/nucleic acids (genetic information)

2 Hydrolysis: splitting of a large molecule (e.g. starch) into smaller molecules (e.g. glucose) by the addition of water. Condensation: joining smaller molecules (e.g. glucose and fructose) into larger molecules (e.g. sucrose) by the removal of water.

3 Soluble molecules can be transported in the blood and across membranes. Insoluble molecules cannot be transported like this, and are excellent storage molecules.

4 Nucleic acids control the functions of cells as they carry the coded genetic information. Proteins regulate the activities of cells e.g. in the role as enzymes. Carbohydrates are the most commonly used source of energy, and keeping cells alive requires energy.

2.8 **1** **a** Fat is solid at room temperature/oil is liquid at room temperature

b In the proportion of these elements – carbohydrates have a higher proportion of oxygen

c Show backbone of glycerol with three fatty acids. Many different fats since there are many different fatty acids.

2 A: washings from a laundry (contain starch); B: milk (contains lactose and protein); C: sweetened tea (gives positive test for sucrose); D: crushed potato (contains starch and small quantities of protein); E: urine from diabetic person (high concentration of glucose)

3 Test that the reagents are working properly by trying the test on a 'known' substance. To show that reagents are not contaminated, carry out the test on water – should have a negative result.

2.9 **1** Source of energy/raw materials for growth and repair/vitamins and minerals to enable other food substances to be used efficiently

2 Autotrophic nutrition involves building organic molecules from simpler inorganic materials e.g. making starch from carbon dioxide and water. In heterotrophic nutrition, the organism obtains the organic molecules ready-made, as food.

3 A balanced diet contains all of the nutrients required for life in the correct proportions

4 Carbohydrates and fats

5 Enzymes (e.g. amylase), hormones (e.g. insulin), structures (e.g. keratin) and transporters (e.g. haemoglobin)

2.10 **1 (p. 45)**

a 8400kJ

b 1000g (10 × 100)

c 18900 − 10500 = 8400 kJ (= 80%)

d i Food 1 as it has the highest energy content per serving

ii 18900/3800 = 4.97 'servings' i.e. 497 g

e 756

1 (p. 47)

a A

b D

c C

d D

e A

2 **a** Colostrum has more protein (400%), less fat (63%), less sugar (44%)

b 1 litre contains 10 × 100cm^3, so contains 20 g of protein

c Any citrus fruit e.g. orange

2.11 **1** Fermentation is respiration by microbes (usually taken to mean anaerobic respiration)

2 To remove any other bacteria which might compete with the 'valuable' bacteria for the nutrients in milk

3

Product	Cheese	SCP	Yoghurt
Starting material	Pasteurised milk	Waste from industrial process	Pasteurised milk
Microbe involved	Lactobacillus	Bacteria	Lactobacillus
Special conditions	Pressure to separate curds from whey	pH, temp etc in bioreactor	Conditions favour lactic acid production, then cool to keep in fridge

4 **a** The enzyme chymosin is made by genetically engineered bacteria

b The design of bioreactors e.g. the development of filters and cooling jackets

5 Different cheeses are made by adding specific bacteria or fungi – the chemicals released by the different microbes gives special flavours to the cheeses. Yoghurt flavour comes from fruit added at the cooling stage.

2.12 **1** **a** Antiseptic kills bacteria outside the body, antibiotics kill bacteria inside the body

b Antibiotic is produced by a mould, antibody is produced by a lymphocyte (type of white blood cell) as a response to an antigen

c Resistance is a biochemical change in a 'target' organism that makes a drug ineffective, immunity involves the action of white blood cells to deal with pathogens

2 It must be able to be digested by digestive enzymes (to release the penicillin)/non-toxic/must not deteriorate with storage/should be cheap to produce

2.13 **1** $C_6H_{12}O_6 \rightarrow 2C_2H_5OH + 2CO_2$

2 Vitamin A: better night vision; vitamin D: better bone growth; calcium: stronger bones and teeth; preservative: so bread does not become infected by fungi/bacteria and turn unattractive to user; whitener: bread may look more attractive to consumer

3 Carbon dioxide: can be sold for use in fizzy drinks, or frozen as 'dry ice' 'spent' (used) hops: very good fertiliser or can be incorporated into animal feed

2.14 **1**

56	20	1120	2489
52	20	1040	2000
54	20	1080	2298

b 10,454; 8,400; 9,652

c 9502 kJ per g. This is more valid since it reduces the effect of any single result.

d Show some means of reducing heat losses to the air, and ensuring that more of heat released by burning raises the temperature of the water. Perhaps burn in oxygen so that peanut burns completely.

e Students repeated experiment using the same set of apparatus e.g. same volume of water/same thermometer. Students converted result to 'per g' to take account of different-sized peanuts.
Students could have taken readings over a period of time and plotted gradient of results to more accurately reflect heat transfer from burning peanut.

2.17 **1** Ethanol dissolves the chlorophyll and removes it from the leaf

2 **a** Light is necessary for photosynthesis

b Carbon dioxide is needed for photosynthesis

c Demonstrates that starch is only produced if chlorophyll is present

d Carbon dioxide, light and chlorophyll

3 **a** Keep plant in dark for 48h – it will be 'destarched', and can be tested for the presence of starch

b Could replace black paper with transparent paper

4 Could heat leaf to denature enzymes – leaf would then not be able to produce starch by photosynthesis

2.18 **1** They are the same size/have the same volume

2 **a** Show curve reaching a plateau at around 7 arbitrary units of light intensity (remember to label the axes)

b 3.5 arbitrary units

c Up to about 7 units – no benefit to grower in exceeding this

d Wavelength of light. Use coloured filters.

e Carbon dioxide concentration/temperature/pH/type of leaf

f To improve validity of data

g To give results as volume produced *per g of plant tissue* i.e. eliminate another potential variable

h Input = temperature, outcome = volume of oxygen produced per minute/fixed = Carbon dioxide concentration/light intensity/wavelength// pH/type of leaf

3 **a** That 'radioactive' CO_2 is treated in the same way as 'normal' CO_2 by the plant

b Otherwise, results obtained in this experiment could not be applied to other plants under 'normal' conditions

c Respiration

2.19 **1** The 1 μm scale is actually 30mm long i.e. 30 000 μm so the magnification of the scale (and the chloroplast) is × 30 000

2.20 **1** Light intensity/temperature/carbon dioxide concentration

2 Can produce acid rain, which is damaging to the mesophyll in leaves. However, higher temperatures and CO_2 concentration can increase the rate of photosynthesis.

3 Magnesium; this ion is a part of the chlorophyll molecule

4 The factor furthest from its optimum i.e. which is actually reducing the rate to the greatest extent.

a Light intensity

b Temperature

c Carbon dioxide concentration

2.21 1 Hydrogencarbonate (bicarbonate) indicator: change from red to purple. Purple colour indicates that solution is less acidic – this is the situation that would apply as CO_2 is removed for photosynthesis.

2 The point at which photosynthesis and respiration are balanced e.g. the amount of oxygen being used in respiration is exactly the same as the amount of oxygen being produced by photosynthesis. This is important since it shows when the plant is beginning to produce more carbohydrate than it is using up i.e. is beginning to make food for the crop.

3 Yes it is true, but it's not the whole story. Plants can use the carbohydrate made during photosynthesis to produce all organic molecules (such as aminao acids and lipids).

4 The plant might take up CO_2 to produce glucose. It would use some of this glucose for respiration, which would mean some CO_2 would be released to the atmosphere. This same CO_2 could then be taken up again as the plant continues to photosynthesise.

2.23 1 Ingestion – Digestion – Absorption – Assimilation – (Excretion/Egestion)

2 a Chemical involves enzymes and digestive juices/mechanical involves teeth, tongue and gut muscles

b Absorption is the transfer of digested foods from the gut to the bloodstream/assimilation is the use of the absorbed food materials in metabolism

c Egestion is the removal of materials which have never been digested and absorbed/excretion is the removal of the waste products of metabolism

3 They prevent the accidental passage of food into the breathing system i.e. the bronchial tree

4 Salivary glands/liver/pancreas/wall of small intestine

2.24 1 It grows, and it certainly is sensitive! The tooth also requires nourishment.

2 Sailors had a very limited diet. It was hard to preserve fruit and vegetables, so the sailors rarely had any foods that contained vitamin C. Without vitamin C the sailors could not make the tough fibres that hold their teeth into their sockets – so their teeth fell out! This is a symptom of scurvy, and was prevented by eating limes or other citrus fruits.

3 b Caries begins when bacteria are able to penetrate the enamel, and infect the dentine of the tooth.

c By avoiding sweet foods (provide nutrients for bacteria, which convert them to acid) and acidic fizzy drinks. Limit caries by regular brushing of teeth to remove food and bacteria, or by adding fluoride to drinking water (strengthens enamel)

2.25 1 a B

b *i* Temperature

ii Beef placed in water

iii To make sure that the beef was sterile – there were no microbes to break down the beef

c Beef in stomach would have been at the right pH, the optimum temperature and so digestion would have been more complete.

2 a Bile

b Emulsifies fats – increases their surface area so that the enzyme lipase can hydrolyse the fats to fatty acids and glycerol more efficiently

3 Mouth: amylase begins to hydrolyse starch to maltose

Stomach: amylase activity stops; protease begins to hydrolyse proteins to peptides under acidic conditions

Duodenum/ileum: amylase hydrolyses more starch to maltose, and maltase hydrolyses maltose to glucose; proteases complete hydrolysis of peptides to amino acids

4 Bile (see question 2); hydrochloric acid in stomach – optimum pH for protease

2.26 **1** There are two inputs – hepatic artery (oxygenated) and hepatic portal vein (digested foods) and one output – hepatic vein (assimilated foods to cells via vena cava)

2 Glycogen and iron are stored. Amino acids are converted to proteins, and excess amino acids are converted to urea

3 Large surface area/lined by thin membrane/good blood supply (capillaries) to keep up concentration gradients by removing absorbed nutrients

4 Goblet cells manufacture and secrete mucus, to protect the lining of the gut from digestive juices. Also found in the lining of the breathing passages, especially the trachea and bronchi.

5 Excess water would be lost in the faeces – diarrhoea

2.27 **1** Low water potential – the dissolved particles tend to 'hang on' to the water molecules, so the water doesn't have much potential to leave

2 Osmosis is the movement of water across a partially permeable membrane down a water potential gradient

3 As cytoplasm swells it presses against the cellulose cell wall. The cell wall stretches slightly, but pushes against nearby cells and so helps to keep the plant body supported (the cells are turgid).

4 **a** Uptake involves enzymes or carriers – the higher temperature gives the moving molecules more kinetic energy

b Energy (from respiration) is needed for this uptake

c This is an example of active transport

d The sugars were used up in respiration to provide the energy needed for the active transport

2.28 **1** Phloem and xylem. Cambium is the dividing tissue.

2 A part of the plant at which a substance is absorbed or manufactured. Leaves are sources of glucose, and roots are sources of water and mineral ions.

3 Sometimes, e.g. towards autumn, the sugar is being stored – may be moving down to roots – or at other times, e.g the spring, it may be move up towards the growing points or flowers

4 To supply energy from respiration – growing points require energy for cell division, roots require energy for active uptake of ions

5 The coloured dye could be placed in a vase/pot and, if it moves up the xylem, it could change the colour of the flower petals

2.30 **1 (p. 94)** 2.4; 4.3; 35.9; 39.0

2 Lower surface is where the most stomata are located – water is lost as water vapour through these stomata, so leaf loses mass

3 So that they were not genetically different, with special adaptations to water conservation

4 Experiment to be repeated and mean values taken. Same amount of Vaseline used each time. Leaves of same size/surface area used.

1 (p. 95)

a Transverse section is cut **across** the leaf

b *i* Photosynthesis
ii Carbon dioxide
iii Diffusion – the movement of molecules down a concentration gradient

c Hairs prevent movement of water vapour/stomata are in pits to trap water vapour/thick waxy cuticle prevents water loss from outer surface

d For photosynthesis/support by turgidity/transport of ions up the xylem

e Sand dunes do not have very much water so it would be hard for the plant to replace any water losses

2 transpiration stream; carbon dioxide; stomata; water vapour; thicker waxy cuticle/leaf rolling/hairs on leaf surface

2.31 **1** Because their surface area to volume ratio is too small for the uptake of gases/loss of wastes. As the organisms get larger they must develop a transport system to move substances to and from the surfaces used for exchange, such as the lungs and kidneys.

2 a Red blood cells and plasma

b Glucose/heat with Benedict's Reagent/positive result would be an orange-red precipitate

c White blood cells

3 Red blood cells: function is transport of oxygen – adapted by absence of nucleus (so very flexible and with large surface area for uptake/loss of oxygen)and presence of haemoglobin (for binding to oxygen).

White blood cells: function in defence against disease – adapted by having irregularly-shaped nucleus (helps cells to squeeze through capillary walls) and cytoplasm which contains digestive enzymes (to break down microbes that have been engulfed).

4 a Jill: many red blood cells as an adaptation to the low oxygen concentrations at high altitude

b Jill: she has a low concentration of the white blood cells needed for defence against disease

c Jackie: low number of platelets, the fragments of cells needed for efficient blood clotting

d Jackie: low number of red blood cells – iron is needed for the production of haemoglobin in red blood cells

e This eliminates any differences in blood cell count due to gender or age – these factors have become fixed variables

5 cells; tissues; epidermis; organ; specialised; red blood cell; division of labour

2.32 1 The skin is a barrier: it has waxy, bacteriocidal ('bacteria-killing') secretions to help prevent water-borne pathogens from crossing into the blood. Pathogens could feed on the body fluids and other tissues, multiplying and causing disease by destruction of cells or release of toxins.

2 a To prevent blood loss/entry of pathogens

b If the clot is internal it might block an important blood vessel and cut off oxygen and nutrient supply to tissues and organs

c *i* It means that a small initial signal can quickly produce a very large response

ii 10 000 000 000 i.e. 10^{10}

3 During phagocytosis, pathogens or harmful particles are engulfed by white blood cells (phagocytes) – the cells form a vacuole around the pathogens (a diagram would help your description!). Phagocytes are very flexible, and contain digestive enzymes to break down the engulfed pathogens. Pathogens can escape phagocytes by 'hiding' inside body cells, or staying in areas where there are no phagocytes (such as in the gut).

2.33 1 By recognising molecules on the surface of the pathogen – these molecules are not found on the cells of the 'host'

2 Lymphocyte has a spherical nucleus, but phagocyte has an irregularly-shaped nucleus. B-lymphocyte makes antibodies, T-lymphocyte kills pathogens directly (e.g. by drilling holes in pathogen's membranes).

3 Pathogen is recognised, and infected animal makes B-lymphocytes to produce antibodies which help to destroy the pathogen. Some B-lymphocytes which can make these specific antibodies are kept in the plasma – they are 'memory cells' – so they are always ready if another attack by the same pathogen takes place.

2.34 1 Arteries have elastic walls with layers of muscles (to cope with pulsing blood pressures, and to continue to push blood towards the tissues): veins have thinner walls (no pulse pressures to deal with) and valves (to prevent backflow of blood)

2 It passes through twice.

Renal vein – vena cava – right atrium – right ventricle – pulmonary artery – (lungs) – pulmonary vein – left atrium – left ventricle – aorta – renal artery

2.35 1 a A:blood, B:tissue fluid, C:lymph

b Oxygen/glucose/amino acids. Carbon dioxide/urea

c Returns along lymph vessel and rejoins blood at thoracic duct close to right atrium (where blood pressure is very low)

d If water potential of blood is high e.g. if concentration of proteins or glucose in blood plasma is very low. Tissues would swell. It can be prevented by making sure that the diet provides sufficient protein.

2.36 1 a Left atrium

b Blood is forced past the semilunar valves and out into the arteries

c Atrium only has to push blood down to ventricles (not very far) but ventricles must push blood out to the tissuesa

2 a By the closing of the valves during the cardiac cycle

b Each beat includes 'lup-dup', so there is one 'lup' per cycle. Each cycle takes 60/72 seconds i.e. there would be 0.8 seconds between successive 'lup' sounds.

2.37 1 By increasing the heart rate (number of beats per minute) and the stroke volume (the amount of blood forced out with each beat)

2 Short-term: heart beats faster and deeper.

Long-term: heart increases in size/power of beat/flexibility of muscle.

3 Higher figure is during the beating of the ventricles, lower figure is during the relaxation of the heart muscle.

2.38 1 a Coronary arteries

b Difficult to move blood past blockage – cells beyond the blockage could be deprived of oxygen/glucose, and so could die

2 a 23 and 46 years of age

b Blood cholesterol is higher in men, and death rate increases as blood cholesterol level increases

c Less than 4 arbitrary units

d There are other factors, as the risk never falls to zero

e Male smoker, who eats a lot of fatty foods and takes little exercise

2.40 1 a B

2 Change of energy from one form to another. The conversion of light energy from the sun into chemical energy in foods is the key step in photosynthesis at the start of food chains. All is eventually released as heat.

3 a Time on x axis, lactic acid concentration on y axis: peak at 25 minutes

b 18 arbitrary units

c 15 minutes

d It is still being made in the muscles and then 'washed' into the blood

e Muscles

f 55 minutes (back to 18 arbitrary units)

2.41 1 a So that they could calculate a mean value

b Seeds: 0.40 mm per s; maggots: 0.52 mm per s. Maggots are more active than seeds, so require some energy for movement.

2 As a control: to show that no 'unknown' factor was responsible for the movement of the coloured liquid

3 a Temperature on x axis, oxygen consumption on y axis: peak at 35°C. This corresponds to the optimum temperature of the enzymes involved in respiration.

b Temperature is the independent/input variable, relative oxygen consumption is dependent/outcome variable

c Mass of organism/surface area of filter paper/type of coloured liquid

4 To remove carbon dioxide from the air before it reaches the living organisms

5 That no carbon dioxide remains in the air

6 Organisms in C have released CO_2. This is an acid gas, and so the indicator changes colour.

7 No organisms in C – this would demonstrate that the indicator doesn't change colour due to some unknown factor

8 Walls of flask may become 'misty': organisms release water from respiration

2.42 **1** Thin, large surface area, close to good blood supply, moist, well-ventilated

2 Along breathing passages (mouth/trachea/bronchus/bronchiole) into air sac. Cross membrane of air sac into capillary and into red blood cell. From red blood cell across wall of capillary and then across cell membrane of e.g. liver cell.

3 Oxygen can be carried in greater quantities because haemoglobin is concentrated in the red blood cells. Red blood cells: function is transport of oxygen – adapted by absence of nucleus (so very flexible and with large surface area for uptake/loss of oxygen)and presence of haemoglobin (for binding to oxygen).

4 Respiration is the release of energy from food molecules. Gas exchange is the transfer of oxygen and carbon dioxide into and out of organisms to allow respiration to go on.

2.43 **1** As these muscles contract they increase the volume of the chest. Lungs follow this change so air is drawn in.

2 Rate of breathing. Volume of air entering lungs at rest and during exercise. Volume of air that can be drawn in in excess of basic breathing. Maximum volume of lungs.

3 Aerobic respiration removes oxygen from inhaled air, and adds carbon dioxide to exhaled air

2.44 **1** Particles trigger the action of phagocytes, and mucus-secreting cells so more mucus is produced. Smoke dries out the lining of the lungs. Chemicals in smoke may trigger extra cell division and start growth of tumours in the lungs or airways.

2 Physical; body cannot function without the chemical. Psychological: user believes that chemical reduces stress or increases pleasure. By offering nicotine patches (to overcome physical addiction) and counselling (to overcome psychological addiction).

3 Increased blood pressure/damage to heart/bladder cancer

4 Cilia are lost and so cannot remove bacteria trapped in mucus

5 **a** 1000

b $6cm^2$

c $6000cm^2$

d $600cm^2$

e Emphysema sufferers have fewer alveoli, so larger spaces/smaller surface areas for gas exchange

2.45 **1** Excretion: the removal of the waste products of metabolism and substances in excess of requirements.

Osmoregulation: control of water balance in the body.

2 Carbon dioxide: respiration in cells – exhalaed from lungs

Urea: deamination in liver cells – removed by the kidneys

3 kidney; renal arteries; nephrons; bowman's; glucose; selective reabsorption; ureters; bladder; urethra

2.46 **1** **a** To keep the cell cytoplasm hydrated

b In food and drink

c In urine, sweat and on the exhaled breath

2 In negative feedback a change in a factor causes a correction that cancels out this change

3 Transplant is more convenient for patient, and action is more natural for the body. Kidney may not be available, and may be rejected by the immune system.

2.47 **1** tissue; receptors; optimum; brain; sensory; negative feedback

2

Organ	Factor controlled
Liver	Blood glucose concentration
Lungs	Concentrations of oxygen and carbon dioxide
Kidney	Urea concentration
Skin	Temperature
Intestines	Availability of digested foods, such as glucose

3 Blood glucose level is detected by the pancreas, insulin is secreted if concentration is too high, liver is stimulated to remove glucose and store it as glycogen, blood glucose level returns to optimum.

Temperature in cabin is measured by thermal sensors, information is sent to on-board computer, electrical message is sent to cabin heaters, heaters are switched on or off to return temperature to optimum.

2.49 1 Similarities: both involved in coordination, both required for survival. Differences: nervous system is much quicker, relies on electrical (not chemical) messages. Nervous system is generally quicker and more localised, which means the effects are useful for short-term survival but not so useful for control of growth.

2 Long (messages carried over great distances), insulated (so no electrical interference), Nodes of Ranvier (allow rapid 'jumping' conduction), many connections with other neurones or muscle cells.

3 a Motor – information out of CNS, sensory – information towards CNS

b Central – brain and spinal cord, peripheral – nerves leading to and from the CNS

4 a As a wave of electrical action potentials

b By neurotransmitter molecules which can diffuse across the synapse

2.50 1 a Stretch receptor (in patellar tendon of the knee). As it stretches it produces an electrical potential

b The sensory neurone

c E (muscle which straightens the knee)

d Total distance covered – there and back – is about 60cm. 100 m = 10 000cm, so this distance would be covered in 60/10 000 s or 6 ms (milliseconds)

e Impulse is slowed as it crosses the synapses

2.51 **1 (p. 148)**

a Cartilage

b It must be elastic, to allow flexion and extension of joints

c To lubricate the joint, and to nourish the cartilage

1 (p. 149)

a Bone is the main structural material – it is very dense, filled with calcium salts, and has no flexibility. Cartilage is softer than bone – some cartilage may become bone if calcium salts are added during growth and development.

b Tendon joins muscle to bone, and does not stretch. Ligament joins bone to bone and does have some ability to stretch.

c When it contracts. The flexor causes bending at a joint e.g. the biceps at the elbow. Extensor causes straightening at the joint e.g. triceps at the elbow.

2 Support: holds up the body against the force of gravity, by providing a framework to attach muscles to.

Protection: bone forms chambers to protect delicate organs e.g. the skull encloses the brain.

Movement: is possible when muscles pull on bones, as long as there are movable joints between the bones. All of these functions are possible because the bone is rigid and incompressible.

3 Calcium and vitamin D. Protein is also important – the bone contains strands of the structural protein collagen – and iron is also important, as the bone marrow is the site of the production of red blood cells (which need iron for haemoglobin).

4 Bending: quadriceps (front of thigh) relaxes and hamstrings (behind thigh) contract.

Straightening: quadriceps contract and hamstrings relax.

2.52 1 a Medulla

b Forebrain (of cerebral cortex)

c Cerebral cortex

2 By the cranium, the bony box of the skull, and by the meninges, the delicate membranes that cover the brain and protect against infection

3 Via sensory neurones, and then out via motor neurones. Many of these neurones arrive through the spinal cord.

4 **a** The 'probe' does not physically enter the body

b The technique can link drug treatment to brain activity, or can locate areas of reduced activity in a damaged brain

2.53 1 **a** 310 ms. The mean is valuable since it reduces the effect of any one result, which might not be a typical result.

b 110–150: 6 students; 160-200: 8 students; 210–250: 10 students; 260–300: 4 students; 310–350: 2 students (did you remember to include student 1?)

c *i* Shortest reaction time is 120 ms, so motorcycle would travel approximately 2m

ii Longest reaction time is 330 ms, so motorcycle would travel approximately 5m

d In a conditioned reflex a second reflex replaces the natural one – here the 'bang' has replaced the need for the subject to observe a light

2.56 1 **a** Stimulant: nicotine (which raises blood pressure); depressant: alcohol (increases reaction times – time to process information); narcotic: heroin (mimics natural painkillers so gives feeling of euphoria)

b Physical; body cannot function without the chemical.
Psychological: user believes that chemical reduces stress or increases pleasure.

c Tolerance: ability of body to function normally even at higher levels of drug. Means that even higher levels will be needed to have an effect on the user.

d Input = caffeine intake in coffee (e.g. number of cups drunk)
Outcome: ability to perform a task involving co-ordination and concentration e.g. recognising different coloured shapes quickly.
Fixed: age/gender/body mass of subjects. Concentration of caffeine in coffee. Time period for drinking coffee.

2 **a** Time on x axis, BAL on y axis. Note peak at one hour.

b *i* 95 mg alcohol per $100 cm^3$ of blood

ii From about 0.6 to 2.9 hours i.e. 2.3 hours

iii Reaction times would be increased e.g. for braking, judgement of distance would be affected

c Skin would appear red

d Carbon dioxide on the breath, water in the urine (and probably as sweat). Long term use causes cirrhosis (hardening of/ loss of function of liver cells).

e Weak bones and teeth, and a tendency to scurvy (bleeding of gums/loss of teeth/poor healing of cuts).

f Shaking of the hands, as muscles contract uncontrollably. Alcohol would stop this, as it is an inhibitor of nervous stimulation of the muscles.

3 **a** *i* The likelihood of addiction

ii Increase = (49.6 – 4.2) = 45.4. Percentage increase = 45.4/4.2 × 100 = 10 800% increase.

iii Increased seizures of cocaine/cannabis resin/LSD. Reduced seizures of heroin. Quantities of LSD are very much lower than of other drugs.

b *i* Suicide/blood infections/malnutrition

ii Methadone gives the sense of euphoria without the physical dependence. Vitamin supplements counteract some of the dangers of malnutrition.

2.57 1 tip; auxin; swell; water; auxin; swell; phototropism; photosynthesis

2 roots; copies/clones; fruiting; pests/weeds; light; water; mineral nutrients; selective; seedless; fertilisation

3.2 **1** **a** 5
b 6
c 7
d 1
e 9
f 4
g 8
h 2
i 3
j 10
2 germinates; flower; reproduction; pollination; fertilisation; fruit; seed; dispersal

3.4 **1** **Pollination** is the transfer of gametes (sex cells) from the anther to the stigma; **fertilisation** is the fusion of these gametes in the ovary.
2 **a** Label sepals and petals (anther and filament wither away)
b The ovary
3 A seed is the part of a fruit that can develop, after germination, into a young plant
4 Tomato – fruit; cucumber – fruit; brussels sprout – neither fruit nor seed; baked bean – seed; runner bean – fruit; celery – neither seed nor fruit; pea – seed; grape – fruit
5 **a** Cover buds with muslin bags so that insects cannot reach flowers. Once flowers are formed, transfer pollen to stigma of half of the available flowers and do not transfer pollen to the other half (if plants are hermaphrodite, cut off anthers to prevent self-pollination). Re-cover flowers with muslin bags. Record which plants produce fruits.
b For self-pollination transfer pollen to stigma of same flower, for cross-pollination transfer to stigma of different flower.
Always make counts of fruits after the same period of time. Keep all plants under same environmental conditions, particularly with respect to soil minerals.

3.5 **1** **a** Should have a positive correlation between time taken and distance travelled
b The longer the time spent floating the greater the distance travelled
c The feathery parachute at the top of the fruit. The fact that the fruit is very light.
d To make these factors into fixed variables – the height could affect the air currents, and the different parts of the laboratory could experience different air currents

3.6 **1** water; micropyle; swell; testa/seed coat; oxygen; respiration; embryo; radicle; positive geotropism; anchorage; plumule; true; photosynthesis
2 Demonstrate that temperature and changes in pH have an effect on the rate of germination

3.7 **1** Sucrose/amino acids
2 Starch (iodine test – blue-black colour is positive result); protein (Biuret test – purple colour is a positive result)
3 Cut off runner and stand in a beaker of red dye – leave for a few hours – cut transverse sections of runner – examine sections under microscope – shoot will have stained red xylem vessels around the outside (root has xylem at the centre)

3.8 **1** **a** To determine the stem length without added auxin
b So that the auxin has had the same time to affect the shoot
c *i* 1mm (5%); 2 mm (10%); 3 mm (15%); 5 mm (25%); 6 mm (30%)
ii Auxin concentration on x axis, percentage change on y axis
iii Percentage change is about 10% so this is a 1 ppm solution
2 **a** To increase the surface area for the absorption of the auxin
b Since shoots are in the dark they cannot photosynthesise so require glucose as a source of energy
c Could show the plant split into two at the level of the small leaves

d Advantage: user knows the exact characteristics of each of the pieces, as they are genetically identical; disadvantage – as they are genetically identical, any disease is likely to affect both of them

e The clear plastic container prevents the loss of too much water (the plantlets have no roots to replace water lost). The container must be clear to allow light to penetrate and allow photosynthesis to take place.

3.9 **1** **a** Allows joining of genetic material from two different parents

b The developmental stage which ensures that the individual has sex organs capable of producing gametes

c This provides haploid cells ready for fertilisation

2 **a** From tubules in testis – along sperm ducts – through prostate gland – along urethra

b From ovary into oviduct

3 Usually, in the first third of the oviduct

4 To replace individuals that have died, and to make sure that there is variation within a sexually-reproducing population

5 **a** R – oviduct, S – vagina

b F in oviduct, I in uterus wall

i Oestrogen/ovary

ii Breasts develop/hips widen/pubic hair develops

6 **a** P – penis, Q – urethra, R – vas deferens/sperm duct

b S to testis, T to testis

c Muscles become stronger, voice becomes deeper, growth of facial hair, more aggressive behaviour

In infected needles/in transfused blood/to baby during childbirth

3.10 **1** Menstruation is the release of the broken down lining of the uterus. Ovulation is the release of an ovum from the ovary. There must be a period of time after ovulation to offer the possibility of fertilisation, before menstruation can take place. Menstruation would include unfertilised ova.

2 FSH – stimulates development of the follicle in the ovary; LH – stimulates release of ovum at ovulation, and development of corpus luteum from remains of follicle; oestrogen – repairs the lining of the uterus; progesterone – keeps the lining of the uterus ready for implantation if fertilisation occurs

3 Progesterone rises in concentration if fertilisation occurs – it then prevents ovulation by 'switching' off the release of FSH, so no more follicles develop. No follicles means no more ova, so no chance of pregnancy.

4 **a** Menstruation – repair – receptive – pre-menstrual

b 28 days

c At about 14 days

3.11 **1** Conception – implantation of the ball of cells formed as the zygote divides; copulation – another name for sexual intercourse, the hard penis is inserted into the vagina; fertilisation – the fusion of the male and female gametes

2 IVF – how many fertilised eggs should be implanted? Should a fertilised egg be implanted in the womb of a surrogate mother? Should an egg donor know who is the sperm donor?

AID – Should a female know who is the sperm donor? Should any attempt be made to match the sperm donor to the female (e.g. for physical characteristics/sporting abilities/intelligence)? When should a child born from AID be made aware of his/her father?

3.12 **1**

Female sterilisation	S
Vasectomy	S
Diaphragm	P
IUD	P
Pill	C
Spermicide	C
Female condom	P
Condom	P

2 Condom or diaphragm – these are temporary methods of contraception

3 Sterilisation is not reversible, and a change of sexual partner might involve a change in the desire for contraception

4 There is so much variation in human physiology, such as body temperature changes, that it is difficult to be sure when a woman has actually ovulated.

3.13 1 The gestation period

2 It takes 42 – 43 divisions

3 Specialisation of cells, which is part of development. Specialisation involves a change in structure and function of cells, and development is a change in the function of an organ.

3.15 1 uterus; oxytocin; progesterone; cervix; amniotic sac; vagina; oxygen; umbilical; placenta; afterbirth

2 Oxytocin – stimulates contractions of uterus; progesterone – concentration falls as labour begins; oestrogen – concentration rises as birth approaches (makes uterus more sensitive to oxytocin). Oxytocin also stimulates lactation, and oestrogen is responsible for the development of the breasts. Human Growth Hormone is one of the controls for growth, especially controlling production of protein in muscles.

3 **a** True

b False

c True

d False

e True

f True

3.16 1 Avoid unprotected sex (for AIDS, gonorrhoea and syphilis), avoid sharing syringes (AIDS), know sexual history of partner (for AIDS, gonorrhoea and syphilis)

2 Should offer testing for STDs, and be aware of treatment available for each type of STD. Should try to trace source of any outbreak, and work out special programmes for those particularly at risk (such as drug users).

3 AIDS (viral), gonorrhoea or syphilis (bacterial)

4 AIDS

Individuals: should know history of sexual partners, avoid unprotected sex

Communities: Should offer testing for AIDS, and be aware of treatment available for AIDS and any associated problems. Should try to trace source of any outbreak, and work out special programmes for those particularly at risk (such as drug users).

Scientists: should devise education programmes to try to limit infection, should try to develop vaccines to prevent infection, should try to develop drugs to manage the disease in people who have become infected.

3.17 1 Genetic background, hormone production, nutrition.

2 How often the measurements should be made (i.e. how quickly the organism is growing). Which parameter is measured.

3 Infancy, childhood, puberty and adolescence (and then adulthood)

3.18 1 (inherited) e.g. eye colour/hair colour/nose shape/presence or ear lobes; (acquired) e.g. body mass/height/any scar

2 Many possible examples – look for relevance and the ability to present the information as a summary which is comprehensible to others in the teaching group

3 **a** A structure, found in the nucleus, made up of DNA and containing the genes. Evidence is that only the chromosomes are transmitted from the sperm, and any damage to the chromosomes tends to cause a change in characteristics.

b They can be observed during cell division – this is because they shorten and coil, making them denser and easier to see.

3.19 1 **a** They are physically damaged as they are forced along blood vessels, and they have no nucleus to control repair processes

b The liver

c Iron (actually Fe^{II}), and it is part of haemoglobin d *i* 5 × 1 000 000 × 5 000 000 = 25 000 000 000 000 i.e. 25 million million

ii About 0.21 million million are replaced every day, so about 2.41 thousand million every second

3.20 **1** **Homologous pair**: chromosomes with the same genes in the same positions – one of the pair comes from the father and one from the mother.
Heterozygous: a nucleus (or organism) carrying both alternative alleles of a gene e.g. Bb.
Homozygous: a nucleus (or organism) carrying only one of the alternative alleles of a gene, although it does have two copies of that allele e.g. bb or BB.

2 A gene is a section of DNA that controls the production of a particular protein, and an allele is an alternative form of a gene

3 A dominant allele is expressed (i.e. is 'in control') in a heterozygous individual. In abbreviated form, the dominant allele is usually shown with a capital letter e.g. B is the dominant allele in the pair Bb.

3.21 **1** Brown eyed parents are both Bb (Brown – B – is dominant to blue – b); gametes from each parent are B or b; at fertilisation one possible combination in the zygote is bb, a blue-eyed child

2 Because the gametes are not produced in exactly equal numbers, and fertilisation is random (i.e. it is not certain that every possible combination will be produced in the expected numbers). The larger the size of the population studied, the closer actual results approach the expectation.

3 This cross enables an experimenter (e.g. an animal breeder) to find out whether an individual showing the dominant characteristic is homozygous or heterozygous. Homozygous individuals might be more valuable, as they will be pure breeding for that characteristic.

3.22 **1** **a** A heterozygote, who has the recessive allele but does not show the recessive characteristic because the dominant allele is also present

b Show parents as Cc, for example. There is a ¼ probability that one of their children will have cystic fibrosis.

2 **a** She was Hh – she showed the condition, but must have had the h allele as she had children who were hh (unaffected)

b B was hh, C was Hh. Show gametes and possible zygotes.

c ½ – remember that each 'event' (i.e. each child) is unaffected by the genotype of its siblings

d Because affected individuals might die before they reproduce, so 'losing' the dominant allele from the population

3 **a** Show parents' genotype, possible gametes, possible combinations in zygotes

b ½

c ¼ : i.e. ½ (female) × ½ (blood group A)

a Codominance

3.23 **1** Show that male produces X and Y, female produces X and X and so there should be equal numbers of XX and XY at fertilisation

2 Males have only one X chromosome, so any allele carried on this X chromosome (which may include the 'colour-blind' allele) will be expressed in the phenotype. A colour blind female must have received the defective allele from both parents: father must be colour blind (X^cY) and mother must be a carrier (X^cX)

3 **a** The mother has the recessive allele, but does not show the condition

b There is a 50% probability that any son will be haemophiliac

3.24 **1** discontinuous; continuous; genes; environmental; genotype; phenotype, phenotype; genotype; environment

2 Blood group

a determined by genes alone;

b all others are affected by environment, such as nutrition

3 **a** Category on *x* axis, number on *y* axis

b Continuous – there are many categories

c Gender/ability to roll tongue

3.25 **1** A gene mutation is an alteration in the DNA on a single chromosome

a the cystic fibrosis allele

b the sickle cell anaemia allele in regions where malaria is common

2 A cancer-causing factor. Some forms of radiation (U-V) can cause skin cancer, chemicals in tobacco smoke can cause lung cancer.

3 Crossing-over during meiosis produces new combinations of alleles in gametes, independent assortment produces new combinations of chromosomes in gametes, and fertilisation has random combinations of gametes from mother and father.

3.26 **1** Adaptation is a change in structure, biochemistry or behaviour which makes an organism more suited to its environment

2 Many possible examples – be careful that you explain how the adaptation gives the organism an advantage in its environment

3.27 **1** **a** 32

b The banded snails are more obvious against the plain background

c *i* Banded (the background is plain)

ii Unbanded (the background has stripes/bands)

d A: snails have many offspring, and there would not be enough food for all of them to live as adults; B: the snails have slightly-different appearances to one another – some have banded shells and some have plain shells; C: the pattern of the banded shells can provide camouflage in a grassy environment; D: the ones with plain shells are seen more easily by predators such as thrushes; E: the snails with banded shells survive longer, breed more often and so pass on their genes to the next generation

3.28 **1** **a** Cut off anthers from 'receiving' flower; collect pollen from 'donor' flower on paintbrush; transfer pollen to stigma of 'receiving' flower. Cover pollinated flowers to prevent pollination by natural means.

b Use vegetative propagation

3.29 **1** 2 – 3 – (1) – 6 – 4 – (8) – 7 – 5

3.30 **1** It can infect plants easily, and carry 'engineered' genes into the host plant

2 New organisms can now fix nitrogen, and so (a) reduce the need for nitrate fertilisers and (b) allow new organism to manufacture more protein

3 Cystic fibrosis: the gene for the 'correct' protein can be carried in a virus into the lungs of a person with the 'faulty' gene. The virus infects the lung lining cells and replaces the faulty gene with the correct one. It is likely to be successful because the virus can be engineered to carry only the correct gene, and none which might be harmful.

4.1 **1** **Population**: all of the members of one species in a particular area; **community**: all of the populations of living organisms in one area; **ecosystem**: all of the living organisms and the non-living factors interacting together in one part of the environment

2 Light intensity/temperature/oxygen concentration/carbon dioxide concentration/humidity/availability of water

3 One may be food for the other (most likely plant being food for animal); insect may be responsible for pollination of flower; animal may help to disperse seeds/fruits of plant

4 Food, shelter and a suitable breeding site

5
- Flying insects are 'cold-blooded' and so cannot remain active at the low temperatures found in Britain during the winter
- During UK winters temperatures are higher in Africa, so there is more active insect life available as food for the swallows
- Hobbies depend on small birds (and large insects for food). Many populations of small birds migrate to Africa in the UK winter as more food is available to them there. The hobbies follow their food.

6 Ecology is the study of living organisms in relation to their environment

7 **a** *i* June

ii April

iii The trees have many leaves, and these absorb the light before it can reach the ground layer

b *i* There is plenty of light for them to photosynthesise before the trees have produced most of their leaves

ii By June the trees are in full leaf and so little light reaches the bluebells. Bluebells cannot photosynthesise without light, and so leaves die back.

4.2 **1 (p. 249)** **Producer**: can trap light energy and produce organic compounds; **consumer**: must obtain its organic food molecules in a ready-made form, from another living organism; **decomposer**: obtain energy and raw materials from the wastes or remains of other living organisms. **Consumers** can be omitted: the cycling of raw materials can go on without consumers but not without producers or decomposers.

2 Whichever example you choose, make sure that there is a producer at the base, and that you show the direction of energy flow with arrows.

3 Energy is lost from each transfer between stages e.g. when a lion eats an antelope, much of the antelope's energy content does not become part of the lion's body (a great deal is lost as heat during the lion's respiration). Because the energy transfer is quite low – only about 5–10% – there cannot be many links to the chain or the number of producers at the base would be enormous.

1 (p. 250)

	Producers	**Herbivores**	**Carnivores**	**Top carnivores**
Freshwater	Algae	Water fleas	Guppy, bass	Green heron
Seashore	Seaweed	Limpet	Crab	Gull
Coral reef	Phytoplankton	Zooplankton, coral, jellyfish	Anemone, parrotfish, sea turtle, butterfly fish, crab, octopus	Shark

2 Many possibilities e.g. number of parrotfish would increase, so coral might be eaten, so less food for butterfly fish, so zooplankton would increase. Main point to make is that food webs are very stable, and removal of any member makes them less stable.

3 energy; producer; light; chemical; herbivore; consumer; carnivore

4.4 **1** bacteria; fungi; simple; amino acids; fats/lipids; enzymes; temperature; pH; sewage; food/buildings

2 **a** They can use it as fertiliser for growing plants

b To provide the moist conditions which supply water for decomposition (hydrolyses in digestion)

c To make sure that oxygen is available, since aerobic respiration by microbes requires oxygen

4.5 **1** **a** Carbon dioxide

b Glucose/starch/cellulose

c Respiration

d Photosynthesis

e Diffusion

f There was no oxygen to complete decomposition/it was too acid for decomposition to take place

4.6 **1** **a** The stubble can be decomposed and release nitrates to the soil

b Drainage increases the air content of the soil, providing oxygen for respiration by microbes. It also raises the temperature of the soil, helping the enzymes of decomposition.

c Peas and beans are legumes – they have a symbiotic relationship with bacteria that can 'fix' nitrogen from the atmosphere into ammonium salts and nitrates, which are excellent fertilisers

d Increases the concentration of nitrate, phosphate and potassium – all mineral ions needed for plant growth

e The compost can be decomposed to release nitrates and other minerals (and the decomposition also raises the soil temperature and moisture content)

2 Drainage increases the air content of the soil, providing oxygen for respiration by microbes. It also raises the temperature of the soil, helping the enzymes of decomposition.

4.7 **1** **a** Herbivore dies – is decomposed – nitrate is released – absorbed by a plant – converted to protein by plant – eaten by herbivore – digested to amino acids – assimilated into herbivore muscles as protein

b Decomposition releases nitrates/ammonium salts for nitrogen cycle, and also releases carbon dioxide for the carbon cycle

2 energy; producer; light; chemical; herbivore; consumer; carnivore; photosynthesis; respiration; decomposers

3 **a** Plants

b Jackals and lions

c Grass > sheep > jackal

d Several jackals can bring down a large sheep; several jackals can search for food; female jackals can remain with pups while males hunt

e The jackals used other species as their prey e.g. dik-diks

f Jugular vein/trachea/spine

g The collars would not be biodegradable i.e. they could not be decomposed by bacteria or fungi.

4.8 **1** Environmental resistance is any factor which limits the growth of a population, the availability of food, for example, or the presence of a predator

2 **Biotic** – 'living': an example of a biotic factor is the presence of a predator; **abiotic** – non-living: an example of an abiotic factor is temperature.

3 To improve yield of crops (e.g. by supplying fertiliser or removing weeds) or domestic animals (e.g by controlling disease)

4.11 **1** This has been done as part of the mechanisation of agriculture: larger fields mean easier harvesting/less habitat for 'pests', for example. This reduces feeding possibilities/nesting sites/hiding places/corridors between habitats for wildlife.

2 Deforestation is the removal of tree cover from land. Humans should be anxious because deforestation upsets water and carbon cycles/causes soil erosion/limits habitat for wildlife (causing extinction of very localised species)

4.13 **1 (p. 274)**

a Year on x axis, population on y axis

b 1400 million

c About 6500 million

d Agricultural, industrial and medical revolutions have combined to increase birth rate and reduce death rate

1 (p. 278)

a Algae > shrimps > fish > sea otter

b The pesticides become more concentrated as they move up the food chain – for example, the otter eats many fish, and so accumulates the pesticide from all of those fish bodies. The pesticide might reach a lethal concentration in the bodies of the otters.

2 **a** Distance downstream on x axis, other variables on y axis. Remember to use a key to distinguish the plots from one another.

b The bacteria had entered in the sewage

c The population falls as the water becomes cloudy with bacteria (no light for photosynthesis) then increases as population of bacteria becomes dispersed and as more nitrates are made available by decomposition of sewage

d The 'rise' in fish numbers is misleading – the data is just showing that the fish are there (they have been there all the time, and haven't appeared from 'zero' fish)

e Describe fall and then increase, giving actual values e.g. fall from 95% to 30% between 0m and 100m from sewage entry

f Fall due to bacterial respiration, then rise as algae photosynthesise (releasing oxygen) and oxygen dissolves in the water from the atmosphere

4.14 **1** Red squirrels are declining in number because of competition with grey squirrels, the fact that the red squirrels are susceptible to a viral disease, and the loss of suitable habitat. A suitable conservation strategy would make sure sufficient habitat was protected, and kept free of invasion by grey squirrels.

2 Management to help species survival. Conservation is necessary because humans are having an impact on other species populations, often by taking habitat from these species.

3 Many suitable examples, but the main point is that there is limited space in the Earth's habitats and humans require some of that space for themselves. The only way that other species can survive is if humans manage the environment for them, although this may benefit some species at the expense of others.

4.17 **1** A biomass fuel is one made from materials produced in photosynthesis

2 They remove CO_2 from the atmosphere, they may limit soil erosion, they may provide a habitat for some animal species, they may reduce the use of other woods for fuel

3 **a** Molasses can be added to cattle feed, dried yeast provides vitamins, and carbon dioxide can be used in fizzy drinks, waste sugar cane can be burned to fuel the distillation process

b Cellulose breaks down plant cell walls and so makes the cell contents accessible. Amylase hydrolyses starch to maltose/glucose, providing the substrate for alcoholic fermentation.

c Genetic engineering has provided strains of yeast which can tolerate the high temperatures and high alcohol concentrations which occur during fermentation

d This suggests that running one car is taking the landspace which could support 25 humans on a subsistence diet. Is this the best use of crop-growing land?

4 Provide cheap source of fuel, and allow communities to get rid of much organic waste

5 The temperature is more stable there, they are easier to fill (gravity), and explosions are less dangerous!

6 The disinfectant would kill some of the active microbes and so slow down the production of biogas

Answers to end-of-topic questions

Characteristics and classification

1. B is 1a/2a/Anopheles; C is 1b/4b/Ornithodorus; D is 1b/4a/5a/Pulex; E is 1a/2b/ Musca; F is 1b/4a/5b/Pediculus
2. **a** Elephas (genus) and maximus (species)
   b *i* Bobcat, European lynx, Iberian lynx OR jaguar, leopard, tiger, lion
   ii Acinonyx
3. **a** Follow key branches to mammal
   b Constant body temperature/vertebral column/different types of teeth
   c Walk upright/communicate by speech and writing/advanced tool use

Cells and organisation

1. Cells: phagocyte, sperm, neurone
   Tissues: epidermis, xylem, blood,
   Organs: liver, heart, leaf, ileum, ovary, brain, stem
2. Organelle – cell – tissue – organ – system - organism
3. Kilometre – metre – millimetre – micrometre;
   There are 1000 micrometres in 1 mm. Therefore number of cells that will fit in = 1000/50 i.e. 20.
4. **a** 41.4
   b Bar chart with appropriate key/labels
   c 13/100 (i.e. 13%) × 100 000 000 000 000 = 13 000 000 000 000
5.

Structure	**Liver cell**	**Palisade cell**
Cell surface membrane	√	√
Chloroplasts	x	√
Cytoplasm	√	√
Cellulose cell wall	x	√
Nucleus	√	√
Starch granule	x	√
Glycogen granule	√	x
Large, permanent vacuole	x	√

6. Choose any example – e.g. neurone (length, number of connections, insulation) – and relate to function – e.g. electrical conduction
7.

Sensitivity	The ability to detect changes in the environment
Respiration	Release of energy from organic molecules
Growth	A change in size
Reproduction	The generation of offspring
Nutrition	The supply of food
Excretion	The removal of the waste products of metabolism
Development	A change in shape or form

8. **a** Chloroplast;
   b Vacuole;
   c Cell wall;
   d Cytoplasm;
   e Chromosomes;
   f Mitochondria;
   g Cell membrane;
   h nucleus
9. Red blood cell – b; white blood cell – g; cell lining bronchiole – f; motor nerve cell – e; palisade cell – i; root hair cell – a; phloem sieve tube – d; sperm cell – c; egg cell - h
10. **a** – organ; **b** – organelle; **c** – organism; **d** – organ; **e** – organism; **f** – organelle; **g** – cell; **h** – organ; **i** – organ; **j** – system; **k** – tissue; **l** – tissue

11 **a** – E; b – A; **c** – D; **d** – C; **e** - B

12 **a** Cells – tissues – epidermis/phloem/xylem – organ – systems – excretory system

b Specialised – red blood cell – division of labour – nervous – endocrine

c Palisade cell – chloroplasts – leaf – epidermis – xylem

Enzymes and biological molecules

1 **a** A protein which is a catalyst in a biological system

b A curve with a peak at pH 2

c *i* In the stomach

ii The rate would fall to zero, as boiling has denatured the enzyme

2 **a** *i* Excretion

ii A protein which acts as a catalyst in a biological system

b *i* pH

ii Temperature/concentration of catalase/concentration of hydrogen peroxide

c 0.57 (divide 17.4 by 10, then take reciprocal of answer)

d Graph drawn with smooth curve and peak at pH 6

e *i* Rate increases to peak at pH 6 and then falls away towards pH 8

ii At higher pH the enzyme is denatured/active site is distorted/less substrate can be worked on/reaction rate will fall as fewer product molecules will be released

3 **a** *i* Mycoprotein has less fat/more fibre

ii Less fat – less risk of blockage to arteries/obesity; more fibre – less risk of constipation/colon cancer

b *i* $100 - (49 + 9.2 + 19.5 + 20.6)\ 100 - 98.3 = 1.7$

ii Mineral such as calcium/iron

c *i* Glucose/amino acids

ii To filter off solid material

iii Carbon dioxide

d *i* 30°C

ii Heat is released during respiration

iii Optimum temperature for the enzymes in the microbes

iv By cool water flowing through the water jacket

4 **a** *i* Graph plotted accurately (peak at pH 8)/axes labelled with correct units included

ii Peak at pH 8 – above and below this pH the enzyme is to some extent denatured, so active site cannot bind so readily to substrate

b *i* Most likely is the subjective nature of end point/also might be inaccuracy in maintaining temperature

ii Use constant temperature water bath/use electronic method of determining when the film had become clear to get valid results. Repeat and take means to improve reliability.

Photosynthesis and plant nutrition

1 **a** *i* Most water is lost from the lower surface of the leaf

ii Upper surface is covered with waxy cuticle, so is almost waterproof, lower surface has stomata for gas exchange, and water may be lost through these

b Leaf was wet to begin with/student had wet fingers when handling cobalt chloride paper

c Repeat the experiment more times

d Compare time taken for cobalt chloride paper to turn pink when attached to lower leaf and when standing 'free' in the atmosphere

2 **a** Osmosis – support – solvent/transport medium – photosynthesis

b Roots – root hairs – surface area – ions/minerals – magnesium – nitrate – active transport

c Xylem – phloem – vascular

3 **a** *i* Carbon dioxide/ water

ii Oxygen

b *i* Iodine solution

ii A, B and D – yellow-brown, C – blue-black

iii B – light is available, but there is no chlorophyll to absorb it/D – neither light nor chlorophyll is available, so no starch is produced

4 a *i* Yellow – yellow – yellow – purple

ii A – respiration releases carbon dioxide, D – photosynthesis removes carbon dioxide

b Pinky-red/photosynthesis and respiration are balanced, so no net change in carbon dioxide concentration.

Animal nutrition and health

1 a *i* Detect starch with iodine solution, which gives a blue-black coloration

ii The product is a reducing sugar – detect reducing sugar by boiling with Benedict's Solution, which gives an orange-red coloration

b *i* Check for accuracy of plot/'dip' corresponding to maximum rate of reaction at pH 4/axes labelled with correct units/key to distinguish between plots

ii Optimum pH at 4.0/presence of salt increases rate of reaction over the whole of the pH range.

c Results will be more **reliable** if the experiments are repeated, and means are taken of collected results. Results will be **valid** if only one variable is changed in any experiment i.e. maintain temperature/concentration of enzyme/concentration of substrate.

2 a Hannah – she has only four fillings but Caitlin has eight

b There may be genetic variation between them/Caitlin may have eaten more sweet, sticky foods, or drunk more fizy, acidic soft drinks

c This removes gender and age from the list of input variables

d The cheek teeth crush food and so the sticky food spends longer there, and can fit into crevices between the teeth

e They may not have had so many cavities, and there would have been fewer teeth in the milk dentition

3

pH	**5.0**	**5.5**	**6.0**	**6.5**	**7.0**	**7.5**	**8.0**	**9.0**
rate	**0.05**	**0.07**	**0.13**	**0.25**	**0.80**	**0.80**	**0.33**	**0.13**

a pH on *x* axis/rate on *y* axis

b Between 7.0 and 7.5

c Ileum: by juices released from the pancreas

d No result with iodine solution/positive result with Benedict's solution

e Temperature/concentration of enzyme/concentration of starch solution

4 a Amylase – Protein – Protease – Fatty acids – Glycerol

b Amino acids are deaminated, the amino group is converted to urea and excreted, the sugar acid is respired/glucose is converted to insoluble glycogen, and stored.

5 Results not valid – might be some other factor affecting healing time, BUT results could be valid, since using only one person meant that many variables were controlled. Improve by using a larger number of subjects.

6

Bounty	818
Maltesers	2016
Mars	1816
Milky way	1960
Minstrels	1775
Snickers	2016
Treets	2476
Twix	2016

b Bar chart – bars should not be touching

c *i* Minstrels

ii Treets

d 1179

e 196.5 minutes; used to maintain temperature/keep nervous sytem working/digest food for example

f 39 minutes

g More – he could use some stored energy e.g. glycogen from the liver

Circulation

1 a For example – oxygen (from lungs to respiring cells); glucose (from intestines to respiring cells/liver)

b 8000 – number can rise during an infection

c 625:1

d Car exhaust fumes/cigarette smoke

e It is reduced – carbon monoxide binds to haemoglobin in place of oxygen

f All red blood cells would have been replaced after 120 days

g The liver stores the iron reclaimed from worn out red blood cells

2 a Aorta

b Pulmonary vein

c Pulmonary artery

d Hepatic vein

e Vena cava (from iliac vein)

f Hepatic portal vein

g Pulmonary artery

3 a It falls

b Blood must be at high pressure to travel through the circulation and form tissue fluid

c Valves

d Muscles in the walls of the small arteries can open or close the vessel so that the flow of blood is controlled

e The capillary walls are very thin, so diffusion is easier, and there are also small pores in the walls of the capillaries

4 a Hepatic portal vein – hepatic vein – vena cava – right atrium – right ventricle – pulmonary artery – pulmonary vein – left atrium – left ventricle – aorta – muscle in leg

b Pulmonary vein – left atrium – left ventricle – aorta – brain – vena cava – right atrium – right ventricle – pulmonary artery

5

rate	55	70	80	90	120	140	150	170
Output per beat	0.07	0.068	0.064	0.068	0.05	0.042	0.038	0.027

b Accurate plot as directed

c *i* Rising to peak then falling,

ii Output per beat falls as rate increases

d *i* There is no benefit in terms of total cardiac output,

ii Output per beat may be higher (heart muscle is more elastic in trained athletes)

e It is less likely that heart muscle will become fatigued/low rate means that there is further to go before the peak rate is reached

f To supply more oxygen to be transported by the blood to the respiring tissues

6 a *i* 5

ii 1.8

b To reduce the number of possible input (independent) variables

7 a Most appropriate would be two overlapping bar charts

b *i* Having blood cholesterol level 45% above normal,

ii Approximately thirty times more likely

8 a Low exercise/smoking
  b High fat/cholesterol/salt
  c Being male!

Gas exchange

1 a Graph plotted appropriately – time on *x* axis
  b Alan – he can get by with fewer breaths during periods of exercise
  c 8:1 (4:0.5)
  d 30 breaths per minute (each one takes 2 seconds)
  e Increases both rate and depth of breathing
  f Heart rate would increase to a maximum as exercise levels increased
  g More oxygen (and glucose) are delivered to the respiring muscles

2 a The diaphragm
  b They will collapse/deflate
  c Exhalation/expiration
  d No intercostal muscles shown
  e 1 – rib; 2 – sternum; 3 – (external) intercostal muscle; 4 – backbone/vertebral column
  f Breathing in involves contraction of the intercostal muscles and the diaphragm, so is 'active'; breathing out occurs when intercostals and diaphragm relax, so is 'passive'

3 a *i* Walking only reduces energy reserves at about 25% of the rate of marathon running
    ii Blood glucose and muscle glycogen
    iii Fat and carbohydrate
    iv 16.6 kJ per g
  b *i* To provide raw material for building muscle
    ii To allow the build up of energy reserves
  c *i* $C_6H_{12}O_6 \rightarrow 2C_3H_6O_3$
    ii The sprint is short, and the harmful effects of lactic acid will not be felt (they would limit the performance of the marathon runner)
    iii Glycogen is broken down, under the influence of glucagon, and then released into the blood.

4 a Excretion/reproduction/movement/sensitivity/growth/nutrition
  b **Respiration** is the release of energy from food molecules. **Breathing** is the set of movements which exchanges the gases oxygen and carbon dioxide.

5 a *i* Cigarette smoke
    ii Cigarette smoke – there is more difference between the numbers for smokers/non-smokers than between the daytime/night time drivers
    iii 8.0 – 5.4 = 2.6, this is because there is less traffic at night, so fewer exhaust gases
  b *i* Nicotine/stimulates heart rate, tar/causes cancer of the lung, sulphur dioxide/damages sense of taste
    ii Reduced birthweight/may be nicotine addict at birth

Excretion and homeostasis

1 a This is a condition in which the body temperature falls significantly below its normal level
  b Elderly people have less efficient circulation, and also their temperature control systems work less well
  c Children have a high surface area to volume ratio, so they lose heat more quickly than they can produce it

2 a Temperature changes are detected by sensors – these then send information to the temperature control centre – messages are sent to the effectors, which try to restore the temperature to its original value
  b Shivering is rapid muscle contractions – these generate heat which raises the body temperature again

c Liver is the main centre of metabolic activity – the many chemical reactions release heat to raise body temperature

d A change from the normal conditions sets off processes which cancel out the change

3 a Show artery connected to aorta, vein to vena cava and tube carrying urine to the entry to the bladder

b Rejection occurs because the body detects that the organ does not belong to the recipient, and the immune system produces more cells which attack the transplant

c The kidney tubule can select which materials will be lost and which ones will be returned from the urine to the blood. The useful materials are returned to the blood by active transport.

4 a It passes by diffusion i.e. down a concentration gradient into the dialysis fluid (the dialysis fluid has a very low urea concentration)

b *i* The transplanted kidney will carry out all of the functions of a normal kidney, and will not run the risk of infection that there is with an external connection to a dialysis machine;

ii The kidney might be rejected. This occurs because the body detects that the organ does not belong to the recipient, and the immune system produces more cells which attack the transplant.

5 a *i* Accuracy of plot/axes labelled with correct units/three lines separated with suitable key

ii A more rapid than B, B slower than C, C more rapid than A

iii A has no covering to reduce heat loss, C loses heat because of evaporation of moisture, B is the best insulated so loses heat least rapidly

b *i* Same sized flasks/same length of time/same volume of water/same starting temperature

ii Stir contents/repeat and take means of results

6 a X – vena cava, Y – ureter, Z – urethra

b Glucose is being reabsorbed/oxygen is being used in respiration/urea is being excreted/water is being reabsorbed and conserved

Receptors and senses

1 A transducer changes one form of energy (the stimulus) into another form of energy (usually an action potential)

a Taste buds convert chemical messages into action potentials

b Rod cells convert light energy into action potentials

2 Cornea – lens – retina – iris – pupil – rods – black and white – cones – colour – retina – inverted – smaller – optic – integration

3 a Ciliary muscles/suspensory ligaments

b Independent is distance from eye, dependent is thickness of lens

c Amount of light in room (could affect pupil size which might have an effect on lens shape)/size and colour of pencil (must make sure that the distance is the only independent variable

d Repeat the measurements at each of the distances, and take a mean of the measurements

4 a A – retina; B – pupil; C – tear duct; D – choroid; E – conjunctiva; F – sclera

b *i* The retina will not be damaged/bleached by the high light intensity;

ii Light – retina – brain – iris muscle

c *i* 20 minutes;

ii The 'two step' shape of the curve suggests that more than one factor is involved

5 a 55 years

b Will be less able to accommodate to objects from different distances, so some images on retina will be blurred

c Once over 60, there is little change in the ability to alter the shape of the lens

d Cornea – accommodation will not be so efficient

e Use model lenses, or lenses removed from animals slaughtered for food. Amount of smoke reaching lens is the independent variable, the measured hardness of the lens is the dependent variable. A suitable control would be a lens not subjected to cigarette smoke (to show that it doesn't harden without the smoke).

6 a *i* D

ii Antagonistic means 'opposite effect' so one muscle contracts as the other relaxes

b *i* Rapid, fixed, does not require conscious thought, has a positive survival value

ii In the lower back region

c *i* Adrenaline

ii Pupils dilate/hairs stand on end/skin goes pale/more glucose is released from glycogen stores/heart beats more quickly

7 a Receptors – Stimuli – Tongue – Nose

b *i* Suspensory ligaments

ii Lens becomes longer and thinner

c *i* 5

ii 2

iii 4

Hormones

1 Adrenalin – Glycogen – Glucose – Oxygen – Deeper – Faster – Gut – Muscles – Pales – Dilate – Stands up – Fight-Flight

2 Male secretes more testosterone at puberty. Testosterone leads to aggressive/ territorial behaviour. As testosterone concentration in blood rises, it inhibits production of further testosterone. As a result, testosterone concentration slowly falls (as it is removed from the blood) and so the male does not become too aggressive.

3 a P – hypothalamus; Q – pituitary gland; R – medulla; S – cerebellum; T – cerebral hemispheres

b *i* Secreted into the blood and then passes to the reproductive organs

ii Child may go through puberty much earlier, or may show greater changes at puberty

Plant reproduction

1 Small – Dull – Light – Dry – Stamens – Style

2 a Anther – produces pollen grains/ sepal – protects flower when in bud/ style – allows passage of pollen tube/ stigma – the surface on which the pollen lands

b Stigma of wind-pollinated is longer/more branched because it must provide a large surface area to catch large quantities of dry/light pollen blowing haphazardly in the wind

3 a Check accuracy/scaling/neatness

b *i* Soft flesh/hard flesh; coloured flesh/white flesh; no stalk/stalk; soft covering to seeds/hard seeds

ii Both contain seeds/both result of sexual reproduction

c Crush flesh/add Benedict's solution/heat in boiling water bath (CARE – wear eye protection)/remove from bath and check for colour change (CARE – avoid skin contact as tube will be hot)/compare colour change – more orange = more reducing sugar.

4 a *i* Pollination is the transfer of pollen from anthers to stigma/ fertilisation is the fusion of male and female gametes

ii Stigma

iii Ovary/ovule

b Seed from fertilised ovule, fruit from ovary

c Wind can be involved in pollination and in fruit/seed dispersal.

5 a *i* For X (200 + 280 + 260 + 260 + 340 + 250 + 240 + 270 + 250 + 290)/10 = 264mm; for Y (400 + 350 + 420 + 610 + 640 + 600 + 340 + 460 + 600 + 520)/10 = 494mm

ii (494 – 264) = 230mm

iii There may be genetic differences, or there may be some environmental factor (e.g. water availability)

b The runners must break, so that the plants are separated

c Asexual reproduction does not allow variation whereas sexual reproduction encourages variation

d *i* As a result of photosynthesis i.e. using light energy to produce glucose from carbon dioxide and water

ii Phloem

Human reproduction and growth

1 a Time on *x* axis/body temperature on *y* axis/menstruation and ovulation indicated

b *i* 36.1 – 36.8;

ii This is to make sure that there are no unexpected independent (input) variables

c *i* Menstruation – fall in temperature, ovulation – increase in temperature;

ii This can indicate the time when conception is most likely (a few days around ovulation) – this time should be avoided for contraception, or represents the best time for intercourse to increase the chance of conception

d The pill/diaphragm both are more reliable as there is much variation in temperature changes during the menstrual cycle

2 a Penis is erected with blood/inserted into vagina

b *i* Barrier

ii Prevents sperm entering the vagina/uterus

c *i* HIV can be transmitted in body fluids, including semen and vaginal secretions

ii In blood transfusions/sharing infected hypodermic syringes/at childbirth

iii The cells which control the action of the immune system are actually infected by HIV

d *i* Leakage of a secretion from the penis

ii Raised temperature

iii Use of antibiotic

3 a It is not losing heat to the environment/it has a high rate of respiration, which generates heat.

b *i* Fat is an excellent insulator against heat loss;

ii This prevents blood reaching the surface of the skin, where heat can be lost to the surroundings

c *i* Change in temperature is detected by sensor – thermostat controls action of heater – more/less heat is released to return temperature to normal;

ii Prevents heat loss to the environment/prevents baby drying out/ allows nurse/mother to view the newborn baby

d *i* Carbohydrate – supplies energy/protein – supplies raw materials for growth;

ii Antibodies – baby may have les immunity to childhood diseases

4 a Age on *x* axis, height on *y* axis/check key for boys and girls

b Weaning at about 1 year/childhood from 2 – 12 years/puberty around 12 – 14 years/ adolescence from 14 – 18 years

c During the first four years

d From 11 – 13 years

Inheritance

1 a *i* Recessive

ii Individuals 6 and 7 do not show the condition, but must be carriers as individual 8 is affected

b There is a 1 in 4 probability – genetic diagram should show two heterozygous parents and indicate the possible allele combinations in their offspring

2 Gene – Meiosis – Diploid – Recessive – Heterozygous

3 **a** The cells become sickled as acid affects the structure of the haemoglobin

b I^NI^S x I^NI^S gives I^N I^S and I^N I^S alleles, and therefore gives I^NI^N, I^NI^S, I^NI^S and I^SI^S offspring

c The heterozygous individuals have red blood cells which are less susceptible to infection by malarial parasites/which are removed more quickly by the liver if they do become infected by malarial parasites.

4 **a** Dominant – Allele – Heterozygote – Genotype

b *i* White

ii Rr x Rr gives alleles R r and R r, so fertilisation provides RR Rr Rr rr

iii 100% red/ 50% red and 50% white

5 **a** *i* TT must be 105

ii Alleles of a gene

iii To simulate the random nature of allele transfer

b *i* Red is dominant, Cross W

ii Flowers from cross V are heterozygous i.e. Tt, so each could provide T or t at gamete formation

c *i* Tt x tt, gametes would be T or t and t or t, offspring would be Tt or tt

ii Cube 1 T T T t t t and cube 2 t t t t t t

Variation

1 **a** Try to breed – no fertile offspring if different species/collect DNA and show differences in DNA fingerprint

b *i* Continuous

ii Equus grevyii

c *i* Phenotype

ii A change in the type or amount of DNA

d *i* Jointed limbs

ii Three parts to body/wings/three pairs of legs

e *i* Horizontal stripes would keep flies away from eyes, ears and mouth

ii Mutation/sexual reproduction/variation/natural selection/increased presence of advantageous mutation

2 **a** In the bones and teeth, as these structures normally contain high concentrations of calcium

b Strontium could cause mutation, which could lead to a tumour in the body cells of the sheep, or could be passed onto successive generations if the mutations are in the gametes.

3 **a** Continuous

b *i* Bars neatly drawn/bars not touching

ii Genes

c *i* A change in the type or the amount of DNA

ii Radiation/chemicals such as those in tobacco smoke

4 **a** *i* 1.2 (9), 1.3 (10)

ii Look for accuracy of bars/bars not touching

b A combination of genetic differences and variation in environment (probably nutrients during development)

Ecosystems, decay and cycles

1 **a** Bottom line: aphid/butterflies/mice; second line: small bird/flies; top line: hawk/spider

b *i* Food web

ii Hawk, spider

c *i* Should be normal pyramid

ii Narrow base – wide mid-bar (aphids) – narrow top bar (small birds)

2 **a** *i* 'odd' pyramid with grass at base, few buffalo, many oxpeckers

ii Pyramid becomes ' normal' pyramid with same organisms in same positions

b **Consumers** obtain their food molecules 'ready-made', by eating other organisms. **Producers** make their own food molecules from simple precursors, using an external energy supply (usually light).

3 **a** O is the absorption of nitrate, released by decomposers and nitrifying bacteria from animal and plant remains, R is the absorption of carbon dioxide to be used in the production of sugars by photosynthesis

b Sulfur dioxide and nitrogen oxides – both may produce acid rain which can lead to damage to leaf tissue/damage to calcium uptake by molluscs and crustaceans/leaching of minerals from soil

4 **a** Peas are legumes so have root nodules with nitrogen-fixing bacteria. This increases nitrate concentration in the soil, so corn grows very well.

b Manure will be decomposed to ammonium compounds and nitrates, which serve as fertiliser for the crops. This is safe and legal as long as there are no pathogens in the animal manure.

5 **a** *i* A – combustion, B – respiration, C – photosynthesis, D – feeding

ii Fungi/bacteria

b More fossil fuels have been burned (to produce more carbon dioxide), and forests have been cut down (so less is removed)

6 **a** *i* Thermal energy

ii Cooling

b *i* Transpiration

ii Windspeed/temperature/humidity

c *i* Fewer trees means less transpiration, so less cloud formation meaning inland areas might become drier

ii Loss of habitat for animals/soil erosion/less removal of carbon dioxide from atmosphere

Human impact on the environment

1 **a** Population increases at entry point of sewage, rises steadily and then falls as distance downstream increases.

b Sewage provides nutrients which allow multiplication of bacteria/ population falls as nutrients are used up.

2 **a** *i* An organism which feeds exclusively on meat

ii The presence of fur

b *i* Fewer antelopes for food/more predation by lions

ii The species could become extinct

c Grass (producer) – antelope (herbivore/primary consumer) – dog (carnivore – secondary consumer) – lion (carnivore – tertiary consumer)

d *i* Human management of ecosystems to maintain populations of other organisms

ii Ensure populations of prey are maintained/limit populations of lions

e Decomposition by fungi/bacteria releases proteins and amino acids/ amino acids are converted to ammonium ions then to nitrates, which can be absorbed by plants.

3 **a** Soft body/shell

b Genus has a capital letter, species has a small/lower case letter

c Sexual/gametes are released and fertilisation occurs

d *i* There will be high oxygen concentrations

ii A change in structure or form during growth

e Any appropriate example – stress management element of conservation

4 **a** Check accuracy

b *i* Sugar beet

ii Wheat – 6 g per m^2

c Improved irrigation/delivery of fertiliser/more efficient harvesting/more efficient storage

d Dry mass excludes water, and water content is very variable

e Each step in a food chain incurs a loss of up to 90% of energy – eating wheat directly eliminates one of the los-making steps

f *i* $6O_2$

ii Large surface area/thin (for diffusion and light penetration)/good transport system/high chlorophyll content of palisade layer *iii* Root cells have a high solute concentration/low water potential, so water enters from soil solution (which has a higher water potential) by osmosis down a water potential gradient.

iv By diffusion through the stomata

5 a *i* Population is mainly made up of young people, with a rapid decline in adults as age increases.

ii The population remains fairly constant throughout all age groups, until more deaths occur above 75 years of age.

b *i* Developing country has more than 20% of total population below 15, developed country has about 6% below 15

ii Developing has less than 5% of population above 65, developed has almost 20% above 65

c There are more females than males above the age of 65

d Female is XX, male is XY so female can produce only X gametes, male produces equal numbers of X and Y. At fertilisation, would expect equal numbers of XX (female) and XY (male) babies.

e *i* Developing = 55 years, developed = 78 years

ii Developed has better medical care/better nutrition/fewer accidents in manual labour/better hygiene (especially supply of clean water)